UNDERGROUND WORKS UNDER SPECIAL CONDITIONS

BALKEMA – Proceedings and Monographs
in Engineering, Water and Earth Sciences

PROCEEDINGS OF THE WORKSHOP (W1) ON UNDERGROUND WORKS UNDER SPECIAL CONDITIONS, MADRID, SPAIN, 6–7 JULY 2007

Underground Works under Special Conditions

Editors

Manuel Romana
Polytechnical University of Valencia, Spain

Áurea Perucho
Laboratorio de Geotecnia (CEDEX), Spain

Claudio Olalla
President of Sociedad Española de Mecánica de Rocas, Spain

CRC Press
Taylor & Francis Group
Boca Raton London New York

CRC Press is an imprint of the
Taylor & Francis Group, an **informa** business

CRC Press
Taylor & Francis Group
6000 Broken Sound Parkway NW, Suite 300
Boca Raton, FL 33487-2742

First issued in hardback 2017

© 2007 by Taylor & Francis Group, London, UK
CRC Press is an imprint of Taylor & Francis Group, an informa business

No claim to original U.S. Government works

ISBN-13: 978-0-4154-5028-7 (pbk)
ISBN-13: 978-1-1384-6578-7 (hbk)

Typeset by Charon Tec Ltd (A Macmillan Company), Chennai, India.

**Visit the Taylor & Francis Web site at
http://www.taylorandfrancis.com**

**and the CRC Press Web site
http://www.crcpress.com**

Underground Works under Special Conditions – Romana, Perucho, Olalla (eds)
© 2007 Taylor & Francis Group, London, ISBN 978-0-415-45028-7

Table of contents

Preface

"*Underground works under special conditions*" is the theme of this Workshop, held in Madrid on the 6th of July 2007, which is organized within the framework of the 11th ISRM International Congress on Rock Mechanics (Lisbon, July 2007).

The question that immediately rises is: do "normal conditions" exist when excavating a tunnel or any other underground work? Tunnelling and mining are almost always hazardous activities, which involve potential dangers and require large doses of good sense engineering and geological awareness. Nevertheless, in the conventional sense, the term "special conditions" implies an increased difficulty, which can often be ascribed to causes of geological, geotechnical and/or technological origin.

This book contains a comprehensive collection of topics related to complex underground works and presents a broad array of problems faced under such conditions. The introduction of this text is formed by the keynote lectures, which (1) provide a clear definition and classification of special conditions that may be encountered (J.A. Hudson) and (2) unveil the latest findings on the Geological Strength Index (GSI) as a valuable tool to deal with rock mass of complicated nature (P. Marinos, V. Marinos and E. Hoek). The technical contributions in this work subsequently present actual findings that resulted from complex underground works that were recently undertaken. They can be grouped in the following themes:

Frequently encountered 'special conditions', with contributions about the Gotthard base Tunnel (Switzerland), the Guadarrama Railway Base Tunnel (Spain), the Cariño fault in an emissary (Spain) and the Maule Tunnel (Chile).

Non-conventional 'special conditions', with contributions treating Ground Reaction Curves and providing insight in creep mechanisms in the Pajares Railway Base Tunnel (Spain).

Advances in technical approaches, such as deformation modulus estimation, an integrated GSI-RMi system, stress variation in longwall supports, numerical simulation of a pre-stressed anchorage and a study of bolts corrosion.

New findings on TBM behaviour, with articles on selection criteria, cutters wear and ground pressures.

Difficult conditions due to the age of the underground work, comprising reports on works on tunnels and a mine and one describing the difficult topographical features op the Telada tunnel (Spain).

The editors sincerely like to thank all authors for their effort. We sincerely hope that this volume will form excellent professional reading and that it will provide significant progress in the engineering of underground works under non-conventional 'special conditions', finally making more possible.

Prof. Manuel Romana
Chairman
Polytechnical University of Valencia
Madrid, Spain, July 2007

Preface

Organization

This Workshop has been organised by the Spanish Society for Rock Mechanics (the ISRM National Group), with the collaboration of CEDEX (the Spanish Official Civil Engineering Research Institution) and of AETOS (the ITA National Group), in the frame of the 11th Congress of ISRM.

Organized by:

11th Congress of ISRM
Workshop W3

in collaboration with:

Centro de Estudios y Experimentación
de Obras Públicas

ORGANIZING COMMITTEE

- Manuel Romana, Spain (Chairman)
- Áurea Perucho, Spain (Co-Chairman)
- Manuel Arnaiz, Spain (President AETOS)
- Celso Lima, Portugal
- Claudio Olalla, Spain
- E. Quintanilha de Menezes, Portugal
- Davor Simic, Spain

SCIENTIFIC COMMITTEE

- Nick Barton, Norway
- Richard Bieniawski, USA
- Benjamín Celada, Spain
- Vicente Cuéllar, Spain
- J. M. Gutiérrez-Manjón, Spain
- Martin Herrenknecht, Germany
- John Hudson, United Kingdom
- Ricardo Lain, Spain
- Paul Marinos, Greece
- Carlos Oteo, Spain
- José Luis Rojo, Spain
- Luis Sopeña, Spain

Keynote lectures

The nature of special conditions in underground construction

J.A. Hudson

Rock Engineering Consultants and Imperial College, UK

ABSTRACT: The nature of the non-standard rock circumstances encountered in underground excavations is highlighted by 12 special conditions: karst geology, problematic environmental situations, adverse fracturing, high rock stress, mixed soil-rock conditions, high water pressure, high temperatures, low temperatures, adverse chemical conditions, proximate structures, unusual project objectives and complex geology. These are described and the ability to model them is tabulated for each case. The necessity to audit computer simulation models incorporating the special conditions is explained because of the associated difficulties. The design of a radioactive waste repository is included as a particularly vivid and perhaps ultimate example of special conditions. The paper concludes on an optimistic note for rock mechanics given the accelerating computing capabilities expected in the years ahead.

1   THE VARIETY OF 'SPECIAL CONDITIONS'

1.1   Introduction

The 'Special Conditions' subject of this Workshop is very timely. We are now at the stage in rock mechanics and rock engineering where we have a large amount of experience. We have superb numerical tools to support the rock engineering design and we have knowledge of how to supply the appropriate input data for the programs in rock engineering construction. Moreover, auditing checks and back analysis can be undertaken to ensure that the design calculations do indeed reflect the rock reality. However, this situation applies to 'standard' projects, and not to 'special' underground conditions, such as the presence of high rock stresses, high or low extremes of temperature as in geothermal energy and refrigerated storage, unusual hydrogeological conditions such as karst geology, severe chemical or biological circumstances affecting rock fractures, etc.

Some of the main 'special conditions' are shown in the 12 squares in Figure 1. These conditions and the associated contexts are summarised in the following paragraphs. It is emphasized that the items have purposely not been placed in any particular order in Figure 1: they can occur individually or in a variety of combinations. However, the items are discussed in clockwise order. Also, the emphasis in this paper is on rock mechanics and rock engineering, rather than also considering soil mechanics and engineering geology issues.

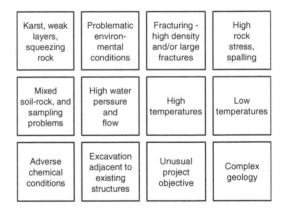

Figure 1.   The variety of potential 'special conditions' that the rock engineer may encounter.

1.2   Adverse rock, karst, weak layers, squeezing

If the rock is weak (or absent in places in the case of a karst formation) or significantly susceptible to water or time effects, there can be problems. In 1977, large solution features were discovered during the construction of the Mark Twain Lake's Clarence Cannon Dam on the Salt River near Hannibal, Missouri, USA. One feature was directly under the dam's left abutment so it needed to be addressed (Fig. 2). All sediments were removed and then the cavities were filled with clay

Figure 2. Large solution features discovered during the construction of the Mark Twain Lake's Clarence Cannon Dam on the Salt River near Hannibal, MO, USA, from Watkins (2006).

Figure 3. Fractured rock mass with large fractures traversing a myriad of small fractures, Mountsorrel granodiorite quarry, UK.

and concrete. A large concrete blanket was placed over the filled cavities.

Sometimes in sedimentary rocks, it is necessary to replace weak rock strata with concrete (Feng & Hudson, 2007). The issue of squeezing problems is one of the topics highlighted for this Workshop.

1.3 Problematic environmental circumstances

An example under this heading is the need to restrict subsidence to an acceptable level beneath houses or other surface structures. Another example is restrictions placed on the extent of the underground works.

One of the environmental aspects of underground construction that will become increasingly important is that of sustainable development. The generally accepted definition of sustainable development is by Bruntland (1987): "Development that meets the needs of the present generation without compromising the ability of future generations to meet their own needs". This leads to the question: Is the exploitation of underground space compatible with the concept of sustainable development? To answer this question for a particular engineering project, we need to consider the factors contributing to the profile of the engineering project noting that:

- underground structures have different functions and lifetimes,
- they have life-spans varying from days (the working face of a longwall coal mine) to thousands of years (radioactive waste disposal),
- they have dimensions varying from metres to kilometres, and

- they have different technological lifetimes, in terms of when they will no longer be required in the form they are currently constructed or envisaged.

Indeed, how does one even approach deciding how to consider the sustainability of underground excavations – either generically or for a specific project? An evaluation framework should be developed so that the profile of a proposed underground structure can be audited for compatibility with sustainable development. So far, this is an essentially untouched area which is ripe for research.

1.4 Adverse rock fracturing – high fracturing density and/or large fractures

The application of rock mechanics principles to structural geology explains the formation of rock fracturing and, because the principal stresses are mutually orthogonal, the fracturing tends to be more regular than might be expected, the rock mass often containing three orthogonal sets. When there have been multiple stages of rock fracturing, the situation becomes more complex.

'Adverse' rock fracturing occurs when there is a high density of fracturing with several fracture sets (Fig. 3) and with weak fracture surfaces. This condition is associated with instability and the need for significant support. An alternative example is the presence of large brittle deformation zones, faults, in crystalline rocks which have little mechanical strength and can be significantly water-bearing.

1.5 High rock stress, spalling

Rock spalling is one of the specific topics to be addressed at the Workshop. Most previous civil engineering projects have not experienced significant rock

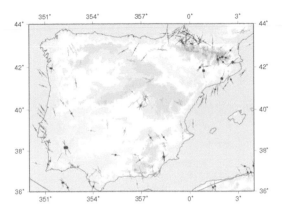

Figure 4. Orientations of the maximum horizontal stress component across Spain and Portugal (from the World Stress Map, Reinecker et al. 2005).

Figure 5. Spalling in the pyrite orebody at the Pyhäsalmi Mine, Finland.

spalling problems because they have been located at relatively shallow depths but, with rock engineering designers intending to penetrate deeper and deeper into the Earth's crust, this will become an increasing issue in underground civil engineering. The subject is familiar to mining engineers because of the deeper nature of some mines and the pioneering work on rock spalling and rockbursts was conducted in South Africa.

The vertical stress component increases by about 1 MPa for every 40 m depth and the maximum horizontal stress is about 2.5 times this value, with the other horizontal stress being intermediate between the two. The high horizontal stress is induced by the mid-Atlantic ridge push causing the overall major principal stress orientation in much of Europe to be NW-SE although, as can be seen from Figure 4, there are significant perturbations to the stress field caused by lithological changes and major faults.

Thus, at 500 m depth, we expect the rock stresses to be of the order of 30, 17.5 and 12.5 MPa. Assuming the Kirsch stress concentration around a circular opening, the maximum concentrated stress will be of the order of $\sigma_{\text{conc}} = 3\sigma_1 - \sigma_3 = 90 - 12.5 = 77.5$. The spalling strength can be about 60% of the UCS so that a rock with a UCS of less than 129 MPa will experience spalling. This stress concentration value will change for non-circular shaped excavations.

An example of rock spalling, in the pyrite ore-body at the Pyhäsalmi Mine, Finland, is shown in Figure 5.

1.6 Mixed soil-rock and sampling problems

Sampling problems arise in weathered and altered rock masses and of course in mixed soil-rock conditions. The case of mixed soil and rock conditions is a difficult one. Another of this Workshop's topics is TBM selection criteria – and the choice of a TBM in mixed

face conditions can be a conundrum. I recall a soft ground bentonite TBM in the UK which was operating in sand within which there were erratic granitic boulders resting on a remanent rock surface. The machine had to be able to simultaneously cope with loose sand and high strength granite!

1.7 High water pressures and flow

In a similar way to the high rock stresses discussed in Section 1.5, the water pressure also increases with depth, although at a rate of 1 MPa/100 m compared to the 1 MPa/40 m for rock, owing to its lesser density.

In most rocks and in the time scales associated with underground construction, water flow is through the rock fractures rather than through the rock matrix. This means that water problems are encountered when the excavation intersects major rock fractures that are connected to the overall fracture network – i.e. at intervals along a tunnel. There have been many examples of extremely high water inrushes into underground excavations, notably the Seikan Tunnel in Japan which was successfully completed after a 17 year construction period in 1988.

1.8 High temperatures

Two examples of underground design having to allow for high temperatures are hot dry rock geothermal energy projects and high level radioactive waste disposal.

In the case of a hot dry rock project, cold water is pumped down one borehole into a region of high rock temperature (Fig. 6). The water flows through fractures in the hot rock mass and is transported back to the surface via another borehole. To design such a project successfully requires an understanding of the

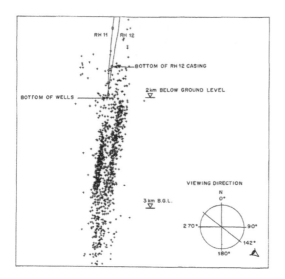

Figure 6. Microseismic emissions recorded during injection of water 2 km deep into the Carnmenellis granite, Cornwall, UK (from Pine & Batchelor 1984).

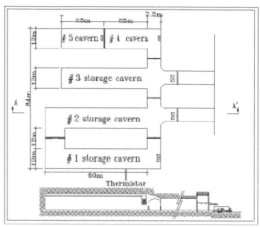

Figure 7. Layout of low temperature food storage caverns near Seoul, South Korea (from Lee et al. 2004).

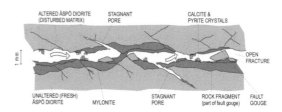

Figure. 8. Consideration of chemical reactions such as dissolution and precipitation that can alter the long-term transmissivity characteristics of rock fractures in crystalline rocks (SKB, 2006).

thermo-hydro-mechanical coupled nature of the system, i.e. in which direction to drill the boreholes, how far apart they should be, etc.

In the case of high level radioactive waste disposal the canisters containing used nuclear fuel emit heat and the rock may typically reach a maximum temperature close to boiling point after 50 to 100 years. This induces an additional stress component which is added to those outlined in Section 1.5 which may enhance the existing potential for spalling and hence affect the groundwater flowpaths and consequently affect the migration of radionuclides to the biosphere.

1.9 Low temperatures

One example of a low temperature underground cavern project is the food storage Gonjiam facility 50 km from Seoul, South Korea (Lee et al. 2004), see Figure 7. The food being stored is mostly frozen meat and fish, and the target operation temperature is −20°C.

After four years of operation, at 1 m from the surface of the cavern wall, the measured temperature was minus 14°C ; whereas, at 10 m from the surface of the cavern wall, the measured temperature was 4°C.

1.10 Adverse chemical conditions

There is a variety of adverse chemical circumstances including strongly acidic conditions, strongly alkaline conditions, and chemicals being transported by groundwater flow through the rock fractures.

In radioactive waste disposal, it is important for the safety analysis to be able to predict the patterns of

groundwater flow for thousands of years. However, if there is significant chemical activity, the hydrogeologically connected fracture network could be considerably altered by chemical processes over the years, in particular precipitation and dissolution effects (Fig. 8).

One of the most dramatic manifestations of dissolution is the karst geology, which was the special condition outlined in Section 1.1. But there are many other circumstances in which chemical activity can be a problem, e.g. corrosion of rock bolts.

1.11 Excavation adjacent to existing structures

This circumstance often has a severe effect on the underground works both through restrictions in the rock mass volume available for the new excavation and through protective measures that have to be taken with regard to both the pre-existing and new structures.

A topical example is the need to protect Antoni Gaudi's Sagrada Familia cathedral in Barcelona during the construction of the rail link from Madrid to Barcelona – which involves the construction of a 12 m

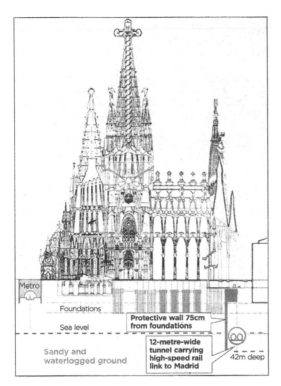

Figure 9. Proposed construction of the underground portion of the Madrid to Barcelona rail link in the immediate proximity of the foundations of the Sagrada Familia cathedral in Barcelona.

Table 1. Uses of underground space for civil engineering purposes.

Civil and military defence, nuclear shelters, aircraft hangars, emergency centres	Refuse management and incineration
Conventional and nuclear power stations	Research facilities, particle accelerators, wind tunnels
District heating	Road tunnels
Drinking water transportation and storage	Shopping malls
Dry docks	Sports, concert, theatre, religious and museum halls
Hot water storage	Storage of food, drink, documents,
Hydroelectric projects	aggregates, minerals
Living quarters	Storage of petroleum products
Radioactive waste storage and disposal	Storm water drainage and storage
Rail transport and stations	Waste water management and treatment plants

Figure 10. Mount Rushmore monumental granite sculptures (18 m high) of U.S. Presidents George Washington, Thomas Jefferson, Theodore Roosevelt, and Abraham Lincoln.

wide tunnel through sandy, water-bearing ground next to the cathedral foundations, Figure 9. A concrete wall is being built to protect the cathedral but there is opposition to the location of the tunnel route at this location.

This special condition of excavation adjacent to existing structures is a major problem in many cities having a long history because of the difficulty in avoiding the plethora of existing underground facilities, and even of knowing where they all are.

1.12 Unusual project objectives

Although there can be many project objectives, as outlined in Table 1, most civil engineering projects involve creating a transport tunnel from A to B, or creating a cavern at a specific location.

However, there can be unusual rock engineering project objectives, a surface example being the Mount Rushmore sculptures of the American presidents, Figure 10.

An example of an artistic underground project is the proposed Chillida sculptural space, or negative sculpture, which will be a 45 m × 50 m × 65 m excavation inside Mount Tindaya on the island of Fuerteventura, Canary Islands. In 1966, the sculptor Eduardo Chillida explained his objective as, "To create a space inside a mountain that would offer men and women of all races and colours a great sculpture dedicated to tolerance…".

The feasibility study is currently being conducted by Arup who note that the artistic requirement for the igneous rock surface to be exposed involves special challenges for the engineering in addition to the engineering and design challenges and that Mount Tindaya presents many environmental and preservation constraints which also dictate the design and construction methods.

1.13 Complex geology

As one example of complex geological conditions, the 5,314 m long Adler tunnel forms part of 'Railway

7

2000' and by-passes the present lines between Basel and Liestal in Switzerland (ITA 2007). The tunnel passes mainly through swelling rock, containing clay and anhydrite. In particular, the Keuper gypsum has a high swelling potential and swelling pressures exceeding 6 MPa, have been measured in laboratory tests under constant volume.

As recorded by the ITA (2007), "The TBM started in November 1995 from the north portal. Over the first 430 metres, several shear zones were encountered. The shale rock in these shear zones was strongly fractured and water-bearing. This led to unstable situations ahead of the cutter face with three cave-ins that propagated to the surface. At the same time the soft and fractured marls became sticky with ground water. The muck stuck to the loading pockets and clogged the openings due to compaction and drying out induced by heat. The hindrance and manual cleaning led to low advance rates".

There are many such instances of 'complex geology' and the subject is relevant to the discussion on modelling difficulties in the next Section. A fundamental problem arises if the geology is too complex to be decoded or if the characteristics of the features cannot be established for modelling purposes.

2 ROCK MASS MODELLING FOR DESIGN PURPOSES – INCORPORATING THE SPECIAL CONDITIONS

2.1 *The ability of computer simulations to adequately incorporate special conditions*

In order to coherently create an underground excavation, it is necessary to have a design, and this design must be based on a model of some kind, whether the model is conceptual, physical, computer simulation, etc. This is because the design of a structure to be contained within a rock mass must be based on a predictive capability and the predictive capability can only be provided by some form of model. If the rock mass cannot be adequately modelled, there is no rational basis for design because it is then not possible to estimate the likely consequences of constructing the structure to any particular design.

With few exceptions, this modelling facility for design is provided nowadays by computer models. Over the last 30 years, computer models have revolutionized our ability to simulate rock masses, especially the presence of rock fractures. However, there are two major problems associated with computer modelling in the context of the 'special conditions' described in this paper.

1. It is not possible to incorporate all the special conditions highlighted in Figure 1; and

Table 2. The capability of computer models for simulating the special conditions.

Special condition	Computer modelling capability
Karst, weak layers	Difficult because of unknown geometry and/or unknown constitutive relations
Environmental issues	Acceptable if a modellable physical consideration, such as subsidence limitation; difficult if a more complex issue such as sustainable development
Rock fracturing	Broad brush treatment OK but detail can be lacking because of the complexity of the fracturing occurrence and the difficulty of obtaining the fracture geometry and mechanical properties
High rock stress	Good if rock stress state is known. Spalling can be well modelled. There are difficulties in establishing the rock stress state and its spatial variation
Mixed soil-rock	Good if the soil-rock boundary geometry and constitutive relations are known
High water pressure	Good providing laminar flow and hydraulic fracturing does not occur
High temperature	Good
Low temperature	Good
Chemical conditions	Difficult because the chemical reactions often occur in fractures and the consequences are not easily linked to standard modelling practices
Proximate excavations	Good
Unusual objective	Good unless it involves special conditions with difficult modelling
Complex geology	Usually not possible because of lack of the necessary detailed information on lithological and fracturing geometry and constitutive behavior.

2. There is currently no auditing procedure in place by which the model and its operation can be checked to ensure that it adequately represents the rock reality.

In Table 2, the current capability of numerical models for simulating the 12 special conditions described in Section 1 is presented. Table 2 indicates not only that there is a mixed capability for computer simulations to adequately represent the rock reality but that there is a consequential need for some kind of auditing procedure to establish to what extent any simulation of rock excavation is valid.

2.2 The need for an auditing procedure to establish the adequacy of modelling

In a paper on a review of techniques, advances and outstanding issues in numerical modelling for rock mechanics and rock engineering prepared by Jing (2003), two text passages in the conclusions are especially important in the current context.

1. "The most important step in numerical modelling is not running the calculations, but the earlier 'conceptualization' of the problem regarding the dominant processes, properties, parameters and perturbations, and their mathematical presentations. The associated modelling component of addressing the uncertainties and estimating their relations to the results is similarly important. The operator should not 'dive in' and just use specific approaches, codes and numerical models, but first consider the specific codes and models to evaluate the harmony between the nature of the problem and the nature of the codes, plus studying the main uncertainties and their potential effects on the results."

2. "Success in numerical modelling for rock mechanics and rock engineering depends almost entirely on the quality of the characterization of the fracture system geometry, physical behaviour of the individual fractures and the interaction between intersecting fractures. Today's numerical modelling capability can handle very large scale and complex equations systems, but the quantitative representation of the physics of fractured rocks remains generally unsatisfactory, although much progress has been made in this direction."

To establish whether a computer simulation is adequate in modelling a rock mass circumstance and whether the special conditions have been sufficiently accommodated, an auditing procedure is required and it is perhaps surprising that in rock engineering this does not yet exist. One can start with some basic questions such as

- What is the work/project objective?
- Have the relevant variables & mechanisms been identified?
- Is the model/code adequate?
- Which data are required?
- How should the data be obtained?
- Are the data adequate?
- Has the model been used properly?
- What are the prediction/back analysis protocols?

However, a more formal auditing approach is required which should be comprised of a 'soft' audit, a 'hard' audit, and an audit evaluation (Hudson et al. 2005) as indicated in Table 3.

Such auditing can be conducted after the modelling has been completed, but it is much better to have contemporaneous auditing so that any necessary

Table 3. 'Soft' and 'hard' audits and audit evaluation.

Soft audit	Hard audit	Audit evaluation
Obtains the information for establishing the essence of the modelling work and provides the ability to present an overview of the modelling	Obtains detailed information on the procedures being used, code development, data selection, etc. with justification of these	Establishes whether the modelling is adequate to meet the objectives, with the associated justification. Discrepancies are noted
⇩	⇩	⇩
Ability to present what is being done	Ability to state the details of what is being done with the justifications	Ability to state whether the modelling is adequate for the purpose

adjustments can be made as the modelling proceeds. A more detailed auditing framework and list of auditing questions is given in Hudson et al. (2005).

Thus, because of the difficulty of some of the special conditions, as noted in Section 1, plus the absence of any currently used auditing procedure for the modelling, we have to conclude that we are currently unable to satisfactorily model all the special conditions encountered in underground construction. In some cases, this lack can be overcome by engineering modifications on site, e.g. via the observational approach, but, in other cases, the situation is more intractable.

2.3 The special conditions of radioactive waste disposal

A particularly vivid, and perhaps the ultimate example of special conditions in underground excavation is the design of a repository for radioactive waste disposal. This is because

- from a technical point of view, the repository can be located anywhere suitable (as opposed to other structures which have to be located in specific places),
- the function of the repository is that of isolation of the waste, i.e. there is no active function, and
- the design life of the repository is measured in hundreds of thousands, if not millions, of years.

Moreover, much more money is available for site investigation (which can amount to 50 km of cored boreholes) than for conventional projects.

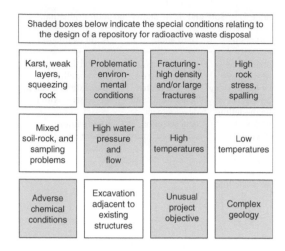

Figure 11. Some of the special conditions associated with designing a repository for radioactive waste disposal.

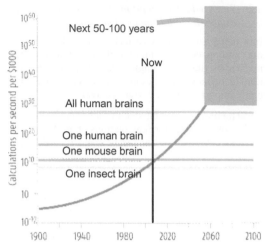

Figure 12. The projected increase in computing power per unit cost from 10^{-1} to 10^{50} calculations/second/\$1000 over the next 100 years from 1900 to 2100, from Kurzweil (2005).

The relevant special conditions in Figure 1 are highlighted in Figure 11 for the case of radioactive waste disposal.

Considering the features highlighted in Figure 11, there are 'problematic environmental circumstances' in terms of locating a repository for hazardous materials. In order to be able to predict the potential migration of radionuclides over a long time period, it is necessary to understand and be able to characterise the hydrogeologically connected fracture network. Such repositories are anticipated at depths of ~400–500 m which means that there will be a high rock stress compared to other civil engineering projects and rock spalling then becomes a possibility. The same applies to the water pressure at these depths. High level waste is significantly heat emitting so that after a period of ~50–100 years the maximum thermal load can induce additional rock stresses of the order of 25 MPa and hence further spalling of the excavations. Given the long design life, it is necessary to be able to model the long-term effects, which then include chemical effects. The project objective is unique: unacceptable quantities of radionuclides should not migrate to the biosphere. Finally, the repository will cover an area of several square kilometres and so it is necessary to understand and characterise all the geological features in the near-field, plus the main features in the far-field.

Needless to say, all these aspects cannot be simulated in one computer model; nor can the results of individual sub-models be validated because of the long design life required. Accordingly, the repository design has to be based on a systems overview methodology (e.g. SKB 2006).

3 THE FUTURE FOR ROCK MECHANICS

The prognosis for successfully modelling the variety of special conditions is not good – for the reasons that have already been outlined. But let us look to the future. What developments can we expect in the decades ahead? In Figure 12, there is a diagram showing the projected growth in computing power expressed as 'calculations per second per \$1000' versus time (Kurzweil 2005). At the moment, for \$1000, one can buy the computing power of one insect brain. In the next 20 years, this price will buy the computing power of one human brain. In the next 50 years, it is anticipated that this price will buy the computing power of all human brains in the world.

We cannot predict where this will lead us: suffice it to say that applications based on computing will grow exponentially in their capability. Certainly, there will be major developments in neural network based applications, possibly leading to some form of machine consciousness.

At the moment, we as human engineers decide that we need a rock engineering facility, then we make preparations, operate a design model, evaluate the consequences, etc. Is it possible that all this could be computerised one day? Futurists say that the last invention that humans need make is the intelligent conscious computer: after that, the computers can take over. Could the computer decide what engineering facilities are required, then indicate what site investigation is required, then choose the most appropriate numerical model from its library, and then decide on the optimal engineering design? This is a long way off

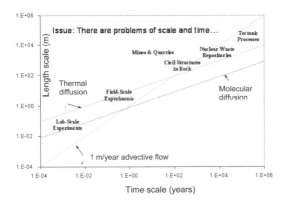

Figure 13. The problems of spatial and temporal scales and extrapolating and interpolating within them is an intractable problem, even with massive computing power (from Yow & Hunt 2002).

but will probably occur at some time in the distant future.

Many of the special conditions may still continue to be intractable in both the short term and long term futures, not because of the computing power required but because of the problem of obtaining the in-situ geometry and mechanical properties of the rock mass components.

The problem of scale is particularly intractable, as illustrated in Figure 13 from Yow and Hunt (2002).

In the short term, we can expect the following developments in rock mechanics and rock engineering.

- Improved methods of accessing/collating information
- More emphasis on geophysical methods in site investigation
- More integration of subjects in computer simulations (e.g. thermo-hydro-mechanical-chemical coupling linked to geology, biology and engineering)
- More international co-operation
- More use of neural network 'intelligent' computer programs
- Larger, deeper and longer excavations
- Emphasis on issues related to the environment, sustainable development, ethical engineering
- Increased rates of mechanised excavation.

Hence, for those of us in rock mechanics and rock engineering, it is an exciting future building larger, longer and deeper excavations with better site investigation, better numerical models, better excavation equipment and much better information and communication facilities. For those young enough to eventually experience the technological developments over the next 30 to 50 years, the future will be even more exciting.

4 SUMMARY AND CONCLUSIONS

1. Attention has been drawn to some of the difficulties potentially encounterable in underground rock engineering through the listing and description of 12 special conditions, highlighted in many of the cases with specific examples.
2. In the majority of underground engineering design cases, it was noted that the design is now established using computer models – some form of model being essential because the design of a structure to be contained within a rock mass must be based on a predictive capability and the predictive capability can only be provided by some form of model. However, the ability of computer models to incorporate the variety of special conditions varies from good to difficult, and sometimes is apparently impossible.
3. This reinforces the notion that some form of formal auditing procedure is required to ensure that the basic computer models, plus those extended to accommodate special conditions, are valid in the sense that they do indeed represent the rock reality. A three-component auditing approach is recommended consisting of a 'soft' audit, a 'hard' audit, and an audit evaluation.
4. The design of a repository for radioactive waste was highlighted as a particularly vivid and perhaps ultimate example of special conditions in the design of an underground rock engineering structure.
5. In the context of the future of rock mechanics, the accelerating development of computing indicates that the ability of computer programs to account for the special conditions may well occur faster than we can currently anticipate. However, this does not obviate the need to solve the problem of obtaining the necessary information regarding the geometry and mechanical characteristics of the rock mass components, which is not simply addressable through enhanced computing power. Furthermore the necessity to interpolate and extrapolate through the spatial and temporal scales will remain intractable in the foreseeable future.

REFERENCES

Bruntland, G. (ed.) 1987. *Our common future: The World Commission on Environment and Development.* Oxford, Oxford University Press.
Hudson, J.A. & Feng, X.T. 2007. Updated flowcharts for rock mechanics modelling and rock engineering design. *Int. J. Rock Mech. Min. Sci.*, 44, 2: 174–195.
Hudson, J.A., Stephansson, O. & Andersson, J. 2005. Guidance on numerical modelling of thermo-hydro-mechanical coupled processes for performance assessment of radioactive waste repositories. *Int. J. Rock Mech. Min. Sci.*, 42, 5–6: 850–870.

ITA,International Tunneling Association, 2007. *http://www.ita-aites.org/cms/373.html.*

Jing, L. 2003. A review of techniques, advances and outstanding issues in numerical modelling for rock mechanics and rock engineering. *Int. J. Rock Mech. Min. Sci.*, 40: 283–353.

Kurzweil, R. 2005. Human 2.0. *New Scientist*, 24 September: 32–37.

Lee, G.S, Jeon, S.W. & Lee, C.I. 2004. Prediction of the temperature distribution around a food storage cavern: the analysis of three-dimensional heat transfer in a fractured rock mass. *Proceedings of the SINOROCK2004 symposium, Int. J. Rock Mech. Min. Sci.*, 41, Supplement 1: 684–689.

Pine, R. J. & Batchelor, A. S. 1984. Downward migration of shearing in jointed rock during hydraulic injections. *Int. J. Rock Mech. Min. Sci.*, 21, 5: 249–263.

Reinecker, J., Heidbach, O., Tingay, M., Sperner, B. & Müller, B. (2005). *The release 2005 of the World Stress Map*. Available online at www.world-stress-map.org.

SKB, Svensk Kärnbränslehantering AB 2006. Long-term safety for KBS-3 repositories at Forsmark and Laxemar – a first evaluation. Main report of the SR-Can project *TR-06-09*.

Watkins, C. 2006. The Meramec Basin Project. *Internet article*. www.rollanet.org/~conorw/cwome/article69&70combined.htm.

Yow, J.L. & Hunt, J.R. 2002. Coupled processes in rock mass performance with emphasis on nuclear waste isolation. *Int. J. Rock Mech. Min. Sci.*, 39, 2: 143–150.

Underground Works under Special Conditions – Romana, Perucho, Olalla (eds)
© *2007 Taylor & Francis Group, London, ISBN 978-0-415-45028-7*

Geological Strength Index (GSI). A characterization tool for assessing engineering properties for rock masses

Paul Marinos
Professor. National Technical University of Athens, School of Civil Engineering, Athens, Greece

Vassilis Marinos
Research Assistant. National Technical University of Athens, School of Civil Engineering, Athens, Greece

Evert Hoek
Consultant, Vancouver, Canada

ABSTRACT: The Geological Strength Index (GSI) is a system of rock-mass characterization that has been developed in engineering rock mechanics to meet the need for reliable input data related to rock-mass properties required as input for numerical analysis or closed form solutions for designing tunnels, slopes or foundations in rocks. The geological character of the rock material, together with the visual assessment of the mass it forms, is used as a direct input for the selection of parameters for the prediction of rock-mass strength and deformability. This approach enables a rock mass to be considered as a mechanical continuum without losing the influence that geology has on its mechanical properties. It also provides a field method for characterizing difficult-to-describe rock masses. Recommendations on the use of GSI are given and, in addition, cases where the GSI is not applicable are discussed.

1 INTRODUCTION

A few decades ago, the tools for designing tunnels started to change. Numerical methods were being developed that offered the promise for much more detailed analysis of difficult underground excavation problems.

Numerical tools available today allow the tunnel designer to analyze progressive failure processes and the sequentially installed reinforcement and support necessary to maintain the stability of the advancing tunnel until the final reinforcing or supporting structure can be installed. However, these numerical tools require reliable input information on the strength and deformation characteristics of the rock mass surrounding the tunnel. As it is practically impossible to determine this information by direct in situ testing (except for back analysis of already constructed tunnels) there was an increased need for estimating the rock-mass properties from the intact rock properties and the characteristics of the discontinuities in the rock mass. This resulted in the development of the rock-mass failure criterion by Hoek and Brown [1980]. A brief history of the development of the Hoek-Brown criterion is to be published in the first issue of a new international journal "Soils and Rocks" [Hoek and Marinos 2007].

The present paper is an update and extension of a paper of Vassilis Marinos et al. [2005].

2 THE GEOLOGICAL STRENGTH INDEX, GSI

Hoek and Brown recognized that a rock-mass failure criterion would have no practical value unless it could be related to geological observations that could be made quickly and easily by an engineering geologist or geologist in the field. They considered developing a new classification system during the evolution of the criterion in the late 1970s but they soon gave up the idea and settled for the already published RMR system. It was appreciated that the RMR system (and the Q system) [Bieniawski 1973, Barton et al. 1974] were developed for the estimation of underground excavation and support, and that they included parameters that are not required for the estimation of rock-mass properties. The groundwater and structural orientation parameters in RMR and the groundwater and stress parameters in Q are dealt with explicitly in effective stress numerical analyses and the incorporation of these parameters into the rock-mass property estimate results is inappropriate. Hence, it was recommended that only the first four parameters of the RMR system (intact rock strength, RQD rating, joint spacing and joint conditions) should be used for the estimation of rock-mass properties, if this system had to be used.

After several years of use it became obvious that the RMR system was difficult to apply to rock masses that are of very poor quality. The relationship between RMR and the constants m and s of the Hoek–Brown failure criterion begins to break down for severely fractured and weak rock masses.

Additionally, since RQD in most of the weak rock masses is essentially zero, it became necessary to consider an alternative classification system. The required system would place greater emphasis on basic geological observations of rock-mass characteristics, reflect the material, its structure and its geological history and would be developed specifically for the estimation of rock mass properties rather than for tunnel reinforcement and support. This new classification, now called GSI, started life in Toronto with engineering geology input from David Wood [Hoek et al. 1992]. The index and its use for the Hoek-Brown failure criterion was further developed by Hoek [1994] and presented in Hoek et al. [1995] and Hoek and Brown [1997] but it was still a hard rock system roughly equivalent to RMR. Since 1998, Evert Hoek and Paul Marinos, dealing with incredibly difficult materials encountered in tunneling in Greece, developed the GSI system to the present form to include poor quality rock masses (Figure 1) [Hoek et al. 1998; Marinos and Hoek 2000 and 2001]. Today GSI continues to evolve as the principal vehicle for geological data input for the Hoek- Brown criterion.

3 FUNCTIONS OF THE GEOLOGICAL STRENGTH INDEX

The heart of the GSI classification is a careful engineering geology description of the rock mass which is essentially qualitative, because it was felt that numbers on joints were largely meaningless for the weak and complex rock masses. Note that the GSI system was never intended as a replacement for RMR or Q as it has no rock-mass reinforcement or support design capability. GSI alone is not a tunnel design tool – its only function is the estimation of rock-mass properties. It is intimately linked at the intact rock strength and should never be used independently of this parameter.

This index is based upon an assessment of the lithology, structure and condition of discontinuity surfaces in the rock mass and it is estimated from visual examination of the rock mass exposed in outcrops, in surface excavations such as road cuts and in tunnel faces and borehole cores. The GSI, by combining the two fundamental parameters of the geological process, the blockiness of the mass and the conditions of discontinuities,

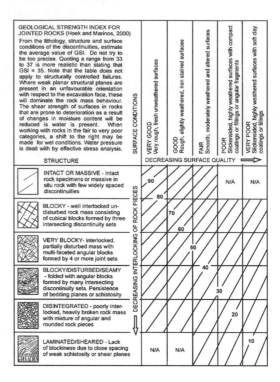

Figure 1. General chart for GSI estimates from the geological observations.

respects the main geological constraints that govern a formation. It is thus a geologically sound index that is simple to apply in the field.

Note that attempts to "quantify" the GSI classification to satisfy the perception that "engineers are happier with numbers" [Cai et al. 2004; Sonmez and Ulusay 1999] are interesting but have to be applied with caution in order not to lose the geologic logic of the GSI system. The quantification processes used are related to the frequency and orientation of discontinuities and are limited to rock masses in which these numbers can easily be measured. These quantifications do not work well in tectonically disturbed rock masses in which the structural fabric has been destroyed. In such rock masses the authors recommend the use of the original qualitative approach based on careful visual observations. Thus, the "quantification" system is only valid in the range of say 35<GSI<75, when the rock mass behavior depends on sliding and rotation of intact rock pieces and where the spacing and condition of discontinuities which separate these pieces and not the intact rock strength control the behavior. When the intact rock pieces themselves can fail then the quantification is no longer valid.

Once a GSI "number" has been decided upon, this number is entered into a set of empirically developed equations to estimate the rock-mass properties which can then be used as input into some form of numerical analysis or closed-form solution. The index is used in conjunction with appropriate values for the unconfined compressive strength of the intact rock σ_{ci} and the petrographic constant m_i, to calculate the mechanical properties of a rock mass, in particular the compressive strength of the rock mass (σ_{cm}) and its deformation modulus (E). Updated values of m_i, can be found in Marinos and Hoek [2000] or in the RocLab program. Basic procedures are explained in Hoek and Brown [1997] but a refinement of the empirical equations and the relationship between the Hoek–Brown and the Mohr–Coulomb criteria have been addressed by Hoek et al. [2002] for appropriate ranges of stress encountered in tunnels and slopes. A recent paper of Hoek and Diederichs [2006] presented new equations for estimating rock mass deformation modulus incorporating measured or estimated intact modulus. These papers and the associated program RocLab can be downloaded from www.rocscience.com.

4 SUGGESTIONS FOR USING GSI

After more than a dozen of years of application of the GSI and its variations for the characterization of the rock mass, this paper attempts to answer questions that have been raised by users about the appropriate selection of the index for various rock masses under various conditions.

4.1 *When Not to Use GSI*

The GSI classification system is based upon the assumption that the rock mass contains a sufficient number of "randomly" oriented discontinuities such that it behaves as a homogeneous isotropic mass. In other words, the behavior of the rock mass is independent of the direction of the applied loads. Therefore, it is clear that the GSI system should not be applied to those rock masses in which there is a clearly defined dominant structural orientation or structurally dependant gravitational instability. However, the Hoek–Brown criterion and the GSI chart can be applied with caution if the failure of such rock masses is not controlled by such anisotropy (e.g., in the case of a slope when the dominant structural discontinuity set dips into the slope and failure occurs through the rock mass). For rock masses with a structure such as that shown in the sixth (last) row of the GSI chart (Figure 1), anisotropy is not a major issue as the difference in the strength of the rock and that of the discontinuities within it is often small. Anisotropy in cases of stress dependant instability is discussed later in this paper.

It is also inappropriate to assign GSI values to excavated faces in strong hard rock with a few discontinuities spaced at distances of similar magnitude to the dimensions of the tunnel or slope under consideration. In such cases the stability of the tunnel or slope will be controlled by the three-dimensional geometry of the intersecting discontinuities and the free faces created by the excavation. Obviously, the GSI classification does not apply to such cases.

4.2 *Geological description in the GSI chart*

In dealing with specific rock masses it is suggested that the selection of the appropriate case in the GSI chart should not be limited to the visual similarity with the sketches of the structure of the rock mass as they appear in the charts. The associated descriptions must also be read carefully, so that the most suitable structure is chosen. The most appropriate case may well lie at some intermediate point between the limited number of sketches or descriptions included in the charts.

4.3 *Projection of GSI Values into the ground*

Outcrops, excavated slopes, tunnel faces and borehole cores are the most common sources of information for the estimation of the GSI value of a rock mass. How should the numbers estimated from these sources be projected or extrapolated into the rock mass behind a slope or ahead of a tunnel?

Outcrops are an extremely valuable source of data in the initial stages of a project but they suffer from the disadvantage that surface relaxation, weathering and/or alteration may have significantly influenced the appearance of the rock-mass components. This disadvantage can be overcome by trial trenches but, unless these are machine excavated to considerable depth, there is no guarantee that the effects of deep weathering will have been eliminated. Judgment is therefore required in order to allow for these weathering and alteration effects in assessing the most probable GSI value at the depth of the proposed excavation.

Excavated slope and tunnel faces are probably the most reliable source of information for GSI estimates provided that these faces are reasonably close to and in the same rock mass as the excavation under investigation.

Borehole cores are the best source of data at depth, but it has to be recognized that it is necessary to extrapolate the one-dimensional information provided by the core to the three-dimensional in situ rock mass. However, this is a problem common to all borehole investigations, and most experienced engineering geologists are comfortable with this extrapolation process.

For stability analysis of a slope, the evaluation is based on the rock mass through which it is anticipated that a potential failure plane could pass. The estimation of GSI values in these cases requires considerable judgment, particularly when the failure plane can pass through several zones of different quality. Mean values may not be appropriate in this case.

For tunnels, the index should be assessed for the volume of rock involved in carrying loads, e.g. for about one diameter around the tunnel in the case of tunnel behavior or more locally in the case of a structure such as the elephant foot of a steel arch. In more general terms the numerical models may include the variability of GSI values over the tunnel in "layers". E Medley and D Zekkos (pers. comm.) are currently considering developing a function defining the variation of GSI with depth for a specific case.

4.4 *Anisotropy*

As discussed above, the Hoek–Brown criterion (and other similar criteria) assumes that the rock mass behave isotropically and that failure does not follow a preferential direction imposed by the orientation of a specific discontinuity or a combination of two or three discontinuities. In these cases, the use of GSI to represent the whole rock mass is meaningless as the failure is governed by the shear strength of these discontinuities and not of the rock mass. Cases, however, where the criterion and the GSI chart can reasonably be used were discussed above.

However, in a numerical analysis involving a single well-defined discontinuity such as a shear zone or fault, it is sometimes appropriate to apply the Hoek–Brown criterion to the overall rock mass and to superimpose the discontinuity as a significantly weaker element. In this case, the GSI value assigned to the rock mass should ignore the single major discontinuity. The properties of this discontinuity may fit the lower portion of the GSI chart or they may require a different approach such as laboratory shear testing of soft clay fillings.

In general terms, when confinement is present, the stress dependant regime is controlled by the anisotropy of the rock masses (e.g., slates, phyllites, etc.). A discussion of anisotropy rock mass behavior in tunneling beyond the commonly used classification systems is presented by Button et al. [2004]. In these cases it would be necessary to develop an orientation dependent GSI. This is a recent idea to try to simplify the treatment of anisotropic problems. However, in view of the potential for complicating the understanding of GSI, an alternative approach may be to use an orientation dependence uniaxial compressive strength. This is more logical from a physical point of view and, being almost completely interchangeable with GSI from a mathematical point of view, should work just as well. The GSI value in this case would be high and the rock mass strength would be determined by the orientation dependant σ_{ci} value.

With the capacity of present day microcomputers, it is also possible to model anisotropy by superimposing a large number of discontinuities on an isotropic rock mass which is assigned a higher GSI value. These discontinuities can be assigned shear strength and stiffness characteristics that simulate the properties of the schistosity, bedding planes and joints in the rock mass. Such models have been found to work well and to give results which compare well with more traditional anisotropic solutions.

4.5 *Aperture of discontinuities*

The strength and deformation characteristics of a rock mass are dependent upon the interlocking of the individual pieces of intact rock that make up the mass. Obviously, the aperture of the discontinuities that separate these individual pieces has an important influence upon the rock-mass properties.

There is no specific reference to the aperture of the discontinuities in the GSI chart but a "disturbance factor" D has been provided in the most recent version of the Hoek–Brown failure criterion [Hoek et al. 2002] and also used in the Hoek and Diederichs [2006] approach for estimating deformation modulus. This factor ranges from D = 0 for undisturbed rock masses, such as those excavated by a tunnel boring machine, to D = 1 for extremely disturbed rock masses such as open pit mine slopes that have been subjected to very heavy production blasting. The factor allows for the disruption of the interlocking of the individual rock pieces as a result of opening of the discontinuities. The influence of this factor is of great significance to the calculated factors of safety.

At this stage (2007) there is relatively little experience in the use of this factor, and it may be necessary to adjust its participation in the equations as more field evidence is accumulated. However, the experience so far suggests that this factor does provide a reasonable estimate of the influence of damage due to stress relaxation or blasting of excavated rock faces. Note that this damage decreases with depth into the rock mass and, in numerical modeling, it is generally appropriate to simulate this decrease by dividing the rock mass into a number of zones with decreasing values of D being applied to successive zones as the distance from the face increases. On the other hand, in very large open pit mine slopes in which blasts can involve many tons of explosives, blast damage has been observed up to 100 m or more behind the excavated slope face. This would be a case for D = 1 and there is a very large reduction in shear strength associated with damage. Hoek and Karzulovic [2000] have given some guidance on the extent of this damage and its impact on rock mass properties. For civil engineering slopes or foundation excavation the blast damage is much more limited in both severity and extent and the value of D is generally low.

This problem becomes less significant in weak and tectonically disturbed rock masses as excavation is generally carried out by "gentle" mechanical means and the amount of surface damage is negligible compared to that which already exists in the rock mass.

4.6 Geological Strength Index at great depth

In hard rock at great depth (e.g., 1,000 m or more) the rock-mass structure is so tight that the mass behavior approaches that of the intact rock. In this case, the GSI value approaches 100 and the application of the GSI system is no longer meaningful.

The failure process that controls the stability of underground excavations under these conditions is dominated by brittle fracture initiation and propagation, which leads to spalling, slabbing and, in extreme cases, rock-bursts. Considerable research effort has been devoted to the study of these brittle fracture processes and a recent paper by Diederichs et al. [2004] provides a useful summary of this work.

When tectonic disturbance is important and persists with depth, these comments do not apply and the GSI charts may be applicable, but should be used with caution.

4.7 Discontinuities with filling materials

The GSI charts can be used to estimate the characteristics of rock-masses with discontinuities with filling materials using the descriptions in the columns of poor or very poor condition of discontinuities. If the filling material is systematic and thick (e.g., more than few cm) or shear zones are present with clayey material then the use of the GSI chart for heterogeneous rock masses (discussed in Heterogeneous and Lithologically Varied or Complex Rock Masses) is recommended.

4.8 The influence of water

The shear strength of the rock mass is reduced by the presence of water in the discontinuities or the filling materials when these are prone to deterioration as a result of changes in moisture content. This is particularly valid in the fair to very poor categories of discontinuities where a shift to the right may be made for wet conditions. The shift to the right is more substantial in the low quality range of rock mass (last rows and columns of the chart).

Water pressure is dealt with by effective stress analysis in design and it is independent of the GSI characterization of the rock mass.

4.9 Weathered rock masses

The GSI values for weathered rock masses are shifted to the right of those of the same rock masses when these are unweathered. If the weathering has penetrated into the intact rock pieces that make up the mass (e.g., in weathered granites) then the constant m_i and the unconfined strength of the σ_{ci} of the Hoek – Brown criterion must also be reduced. If the weathering has penetrated the rock to the extent that the discontinuities and the structure have been lost, then the rock mass must be assessed as a soil and the GSI system no longer applies.

4.10 Heterogeneous and lithologically varied or complex rock masses

GSI has been extended to accommodate the most variable of rock masses, including extremely poor quality sheared rock masses of weak schistose materials (such as siltstones, clay shales or phyllites) often inter-bedded with strong rock (such as sandstones, limestones or quartzites). A GSI chart for flysch, a typical heterogeneous lithological formation with tectonic disturbance, was published by Marinos and Hoek [2001]. This chart has recently been revised (Marinos unpubl.) and the revised version is reproduced in Figure 2. This revision is based on recent experience from a number of tunnels constructed in Greece. It includes cases of siltstones with little disturbance and a variety of cases of siltstones alternating with good rock (e.g., sandstone).

For lithologically varied but tectonically undisturbed rock masses, such as the molasses, a new GSI chart was presented in 2005 [Hoek et al. 2005] (Figure 3). For example, molasse consists of a series of tectonically undisturbed sediments of sandstones, conglomerates, siltstones and marls, produced by the erosion of mountain ranges after the final phase of an orogeny. They behave quite differently from flysch which has the same composition but which was tectonically disturbed during the orogeny. The molasses behave as continuous rock masses when they are confined at depth and the bedding planes do not appear as clearly defined discontinuity surfaces. Close to the surface the layering of the formations is discernible and only then similarities may exist with the structure of some types of flysch.

In design the intact rock properties σ_{ci} and the m_i must be also considered. A "weighted average" of the properties of the strong and weak layers should be used.

In a recent publication [Marinos et al. 2006] a quantitative description, using GSI, is presented for rock masses within an ophiolitic complex. Included are types with large variability due to their range of petrographic types, their tectonic deformation and their alternation (Figure 4). The structure of the various masses include types from massive strong to sheared weak, while the conditions of discontinuities are in most cases fair to very poor due to the fact that they are affected by serpentinisation and shearing. This description allows the estimation of the range of

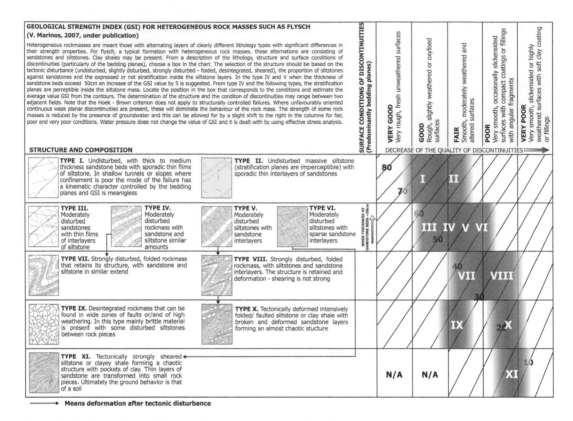

Figure 2. Geological Strength Index for heterogeneous rocks such as flysch.

properties and the understanding of the dramatic change in tunneling, from stable conditions to severe squeezing within the same formation at the same depth.

4.11 *Rocks of low strength of recent age*

When rocks such as marls, claystones, siltstones and weak sandstones are developed in geologically stable conditions in a post tectonic environment, they usually present a simple structure with no or few discontinuities.

When these rocks form continuous masses with no discontinuities, the rock mass can be treated as intact with engineering parameters given directly by laboratory testing. In such cases the GSI classification is not applicable.

In cases where discontinuities are present, the use of the GSI chart for the "blocky" or "massive" rock masses (Figure 1) may be applicable. The discontinuities in such weak rocks, although they are limited in number, cannot be better than fair (usually fair or poor) and hence the GSI values tend to be in the range of 45–65. In these cases, the low strength of the rock mass results from low intact strength σ_{ci}.

5 PRECISION OF THE GSI CLASSIFICATION SYSTEM

The "qualitative" GSI system works well for engineering geologists since it is consistent with their experience in describing rocks and rock masses during logging and mapping. In some cases, engineers tend to be uncomfortable with the system because it does not contain parameters that can be measured in order to improve the precision of the estimated GSI value.

The authors do not share this concern as they feel that it is not meaningful to attempt to assign a precise number to the GSI value for a typical rock mass. In all but the very simplest of cases, GSI is best described by assigning it a range of values. For analytical purposes this range may be defined by a normal distribution with mean and standard deviation values assigned on the basis of common sense. GSI, with its

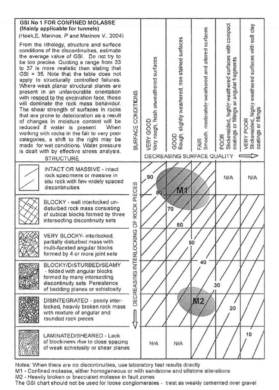

Notes: When there are no discontinuities, use laboratory test results directly
M1 - Confined molasse, either homogeneous or with sandstone and siltstone alterations
M2 - Heavily broken or brecciated molasse in fault zones
The GSI chart should not be used for loose conglomerates - treat as weakly cemented river gravel

Figure 3. Chart for confined molasse (mainly applicable for tunnels).

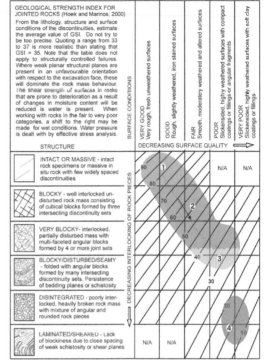

1. Massive strong peridotite with widely spaced discontinuities. The conditions of discontinuities are poorly only affected by serpentinisation
2. Good to fair quality peridotite or compact serpentinite with discontinuities which may be severely affected from alteration.
3. Schistose serpentinite. Schistosity may be more or less pronounced and their planes altered.
4. Poor to very poor quality sheared serpentinite. The fragments are also consisting from weak materials
⟶ Increase of presence of serpentines or other weak material (e.g talc) in joints or schistosity

Warning: The shaded areas indicate the ranges of GSI most likely to occur in these type of rocks. They may not be appropriate for a particular site specific case.

Figure 4. Ranges of GSI for various qualities of peridotite-serpentinite rock masses in ophiolites.

qualitative principles of geological descriptions, is not restrained by the absence of good exposures or the limitations of quantitative core descriptions.

Although GSI is a totally independent system, in the earlier period of its application it was proposed that correlation of "adjusted" RMR and Q values with GSI be used for providing the necessary input for the Hoek – Brown criterion. Although this procedure may work with the better quality rock masses, it is unreliable in the range of weak (e.g., GSI<35), very weak and heterogeneous rock masses, where these correlations are not recommended.

Whenever GSI is used, a direct assessment, based on the principles and charts presented above, is recommended. Fortunately most GSI users have no difficulty in thinking of it as a totally independent system. However, in cases of comparisons or back analysis where other classification systems have been used, some kind of correlation with these other systems is needed. In such cases it may be useful to consult a recent paper [Tzamos and Sofianos 2007]. The four classification-characterization systems (RMR, Q, RMi [Palmström 1996] and GSI) were investigated and all systems rating are grouped in a common

fabric index chart. The reader is reminded not to lose sight of the real geological world in considering such correlations.

6 GSI AND CONTRACT DOCUMENTS

One of the most important contractual problems in rock construction and particularly in tunneling is the issue of "changed ground conditions". There are invariably arguments between the owner and the contractor on the nature of the ground specified in the contract and that actually encountered during construction. In order to overcome this problem there has been a tendency to specify the anticipated conditions in terms of tunneling classifications. More recently some contracts have used the GSI classification for this purpose, and the authors are strongly opposed to this trend.

As discussed earlier in this paper, GSI was developed solely for the purpose of estimating rock-mass properties. Therefore, GSI is only one element in a tunnel design process and cannot be used, on its own, to specify tunneling conditions. It has to be associated with the intact rock strength, with the petrographic constant m_i and with all the characteristics (such as anisotropy) of the rock mass that may impose a different mode of failure than that of a stressed homogeneous isotropic rock mass.

The use of any classification system to specify anticipated tunneling conditions is always a problem as these systems are open to a variety of interpretations, depending upon the experience and level of conservatism of the observer. This can result in significant "changes" in excavation or support type and can have important financial consequences.

The geotechnical baseline report [Essex 1997] was introduced in an attempt to overcome some of the difficulties and has attracted an increasing amount of international attention in tunneling.

7 CONCLUSIONS

Rock-mass characterization has an important role, not only to define a conceptual model of the site geology, but also for the quantification needed for analyses "to ensure that the idealization (for modeling) does not misinterpret actuality" [Knill 2003]. If it is carried out in conjunction with numerical modeling, rock-mass characterization presents the prospect of a far better understanding of the mechanics of rock-mass behavior [Chandler et al. 2004]. The GSI system has considerable potential for use in rock engineering because it permits many characteristics of a rock mass to be quantified thereby enhancing geological logic and reducing engineering uncertainty. Its use allows the influence of variables, which make up a rock mass, to be assessed and hence the behavior of rock masses to be explained more clearly. One of the advantages of the index is that the geological reasoning it embodies allows adjustments of its ratings to cover a wide range of rock masses and conditions but it also allows us to understand the limits of its application.

ACKNOWLEDGEMENT

The paper is the result of a project co-funded by the European Social Fund (75%) and Greek National Resources (25%) – Operational Program for Educational and Vocational Training II (EPEAEK II) and particularly the Program PYTHAGORAS.

REFERENCES

Barton NR, Lien R, Lunde J [1974]. Engineering classification of rock masses for the design of tunnel support. Rock Mech 6(4):189–239.

Bieniawski ZT [1973]. Engineering classification of jointed rock masses. Trans S Afr Inst Civ Eng 15:335–344.

Button E, Riedmueller G, Schubert W, Klima K, Medley E [2004]. Tunnelling in tectonic mélanges-accommodating the impacts of geomechanical complexities and anisotropic rock mass fabrics. Bull Eng Geol Env 63:109–117.

Cai M, Kaiser PK, Uno H, Tasaka Y, Minami M [2004]. Estimation of rock mass strength and deformation modulus of jointed hard rock masses using the GSI system. Int J Rock Mech Min Sci 41(1):3–19.

Chandler RJ, de Freitas MH, Marinos P [2004]. Geotechnical characterization of soils and rocks: a geological perspective. In: Proceedings of the Advances in Geotechnical Engineering: The Skempton Conference, vol 1, Thomas Telford, London, pp 67–102.

Diederichs MS, Kaiser PK, Eberhardt E [2004]. Damage initiation and propagation in hard rock during tunneling and the influence of near-face stress rotation. Int J Rock Mech Min Sci 41(5):785–812.

Essex RJ. [1997]. Geotechnical baseline reports for underground construction. American Society of Civil Engineers, Reston.

Hoek E [1994]. Strength of rock and rock masses. News J ISRM 2(2):4–16.

Hoek E, Brown ET [1980]. Underground excavations in rock. Institution of Mining and Metallurgy, London.

Hoek E, Brown ET [1997]. Practical estimates of rock mass strength. Int J Rock Mech Min Sci Geomech Abstr 34:1165–1186.

Hoek E, Diederichs MS [2006]. Empirical estimation of rock mass modulus. Int J Rock Mech Min Sci 43:203–215.

Hoek E, Karzulovic A [2000]. Rock mass properties for surface mines. In: Hustralid WA, McCarter MK, van Zyl DJA, eds. Slope stability in surface mining. Society for Mining, Metallurgical and Exploration (SME), Littleton, pp. 59–70.

Hoek E, Wood D, Shah S [1992]. A modified Hoek–Brown criterion for jointed rock masses. In: Hudson JA ed. Proceedings of the rock mechanic symposium. "International Society of Rock Mechanics Eurock" 92, British Geotechnical Society, London, pp. 209–214.

Hoek E, Kaiser PK, Bawden WF [1995]. Support of underground excavations in hard rock. AA Balkema, Rotterdam.

Hoek E, Marinos P [2007]. A brief history of the development of the Hoek-Brown failure criterion. Intern. Journ. Soils and Rocks, in print.

Hoek E, Marinos P, Marinos V [2005]. Characterization and engineering properties of tectonically undisturbed but lithologically varied sedimentary rock masses under publication. Int J Rock Mech Min Sci 42:277–285.

Hoek E, Marinos P, Benissi M [1998]. Applicability of the geological strength index (GSI) classification for weak and sheared rock masses – the case of the Athens schist formation. Bull Eng Geol Env 57(2):151–160.

Hoek E, Caranza-Torres CT, Corkum B [2002]. Hoek–Brown failure criterion-2002 edition. In: Bawden HRW, Curran J, Telsenicki M, eds. Proceedings of the North American

Rock Mechanics Society (NARMS-TAC 2002). Mining Innovation and Technology, Toronto, pp. 267–273

Knill J [2003]. Core values (1st Hans-Closs lecture). Bull Eng Geol Env 62:1–34.

Marinos P, Hoek E [2000]. GSI: a geologically friendly tool for rock mass strength estimation. In: Proceedings of the GeoEng2000 at the International Conference on Geotechnical and Geological Engineering, Melbourne, Technomic publishers, Lancaster, pp. 1422–1446.

Marinos P, Hoek E [2001]. Estimating the geotechnical properties of heterogeneous rock masses such as flysch. Bull Eng Geol Env 60:82–92.

Marinos P, Hoek E, Marinos V [2006]. Variability of the engineering properties of rock masses quantified by the geological strength index: the case of ophiolites with special emphasis on tunnelling. Bull Eng Geol Env 65: 129–142.

Marinos V, Marinos P, Hoek E [2005]. The geological Strength index: applications and limitations. Bull Eng Geol Environ 64:55–65.

Palmström A. [1996]. Characterizing rock masses by the RMi for use in practical rock engineering, part 1: the development of the Rock Mass Index (RMi). Tunnell Undergr Space Tech 11(2):175–88.

Sonmez H, Ulusay R. [1999]. Modifications to the geological strength index (GSI) and their applicability to the stability of slopes. Int J Rock Mech Min Sci 36:743–760.

Tzamos S, Sofianos AI [2007]. A correlation of four rock mass classification systems through their fabric indices. Int J Rock Mech Min Sci. in press.

Technical papers

Hard rock bursting phenomena in Maule tunnel (Chile)

Fernando Abadía Anadón

DRAGADOS S.A. Technical Department, Underground and Water Treatment Works

ABSTRACT: Maule tunnel is a part of the hydroelectric complex Pehuenche, located in the seventh region of Chile, some 67 Km to the east of Talca town and some 250 Km to the south of Santiago city.

It is aimed at diverting a flow volume of $140\,m^3/s$ from Maule river to Melado dam in order to feed Pehuenche hydroelectric power station of 500 MW.

The tunnel is 6,755 m long and its cross section is a hydraulic horseshoe of $53\,m^2$, 8 m in equivalent diameter.

The tunnel run mainly across grain-diorites, usually hard an well crystallized, apart from some 700 m at Maule mouth where the ground was chiefly made up of andesites. Its overburden reached 1,300 m.

When the tunnel overburden exceeded 900 m, hard rock bursting phenomena sprang out. It meant that 1,275 m of tunnel had to be excavated under these conditions.

The present paper is aimed at describing the rock bursting phenomena, the troubles that this phenomenon inflicts on normal drill and blast routine and the procedures assumed to afford the tunnel excavation under these conditions.

1   INTRODUCTION

Maule tunnel is a part of the hydroelectric complex Pehuenche, located in the seventh region in Chile, some 67 Km to the east of Talca town and some 250 Km to the south of Santiago city.

It is aimed at diverting a flow volume of $140\,m^3/s$ from Maule river to Melado dam in order to feed Pehuenche 500 Mw hydroelectric power station taking back the water to the Colbum reservoir tail.

The tunnel is 6,755 m long and its cross section is a hydraulic horseshoe of $53\,m^2$, 8 m in equivalent diameter. Its slope reaches 2.3 ‰ in decline from Maule river to Melado dam.

The tunnel run mainly across grain-diorites, usually hard an well crystallized, apart from some 700 m at Maule mouth where the ground was chiefly made up of andesites. Its overburden reached 1,300 m.

2   GEOLOGY

The tunnel geological background before its excavation was founded on seismic refraction profiles, test drillings and two exploratory galleries, one in each portal.

There were several parallel refraction profiles on each tunnel inlet. All them showed a high velocity horizon in depth corresponding to a sound rock with good features.

At Maule portal the samples showed the ground was made up of andesites and tufas displaying admissions between 3 and 10 Lugeon units and being the highest permeability at the surface levels. At Melado portal the test drillings showed grain-diorites and the admissions were low, between 0.5 and 3 Lugeon units.

There were two research galleries, one in each portal. At the entrance portal (Maule) there was a gallery, over 600 meters in length, located 9 meters above the tunnel top. In the first section, some 140 meters made up of andesites, the rock presented overall medium fracture, but intense at some sections, displaying scarce and small seepages. In the second section, up to 473 meters, the rock quality improved and the maximum quoted seepages reached a flow volume of 3.16 litres per second.

At the exit portal (Melado) there also was around a short gallery fully in grain-diorites showing an excellent geotechnical quality, there was only weak leakages from the gallery roof.

The geological tunnel profile (figure 1) pointed out normal contacts between the grain-diorites and the andesites, without slickenside. There only was some photo-alignments that they could be understood as a fault section.

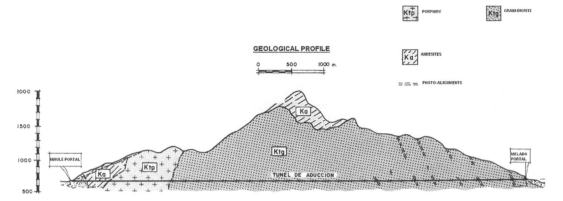

GEOLOGICAL PROFILE

0   500   1000 m.

Figure 1.

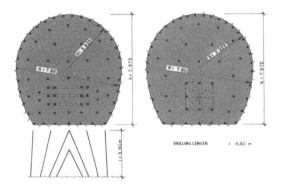

Figure 2.

3 THE PROJECT

According to the previous report the project foresaw a tunnel made by drill and blast methods and lined with shotcrete, occasionally with wire mesh, and systematic bolts 3 meters long.

Possibly some metallic ribs would be used and solely was deemed necessary to line the 12% of the tunnel length with conventional plain concrete. The stretches would be defined along the excavation process.

In the wake of these data it was decided to tackle the tunnel by digging it in full section, excavating simultaneously from both portals, the Maule and Melado one, and making use of on-track equipment. Drilling was entrusted to two three-boomed Jumbos endowed with hydraulic hammers.

Two drilling schemes have been mainly used (figure 2), the parallel and the wedge one. The parallel one was used to drill 4.5 meters long blast in sections without troubles, and the wedge one was used to drill either three meters or shorter blasts for sections with geotechnical troubles.

4 ROCK BURSTING WARNING PARAMETERS

At the time the tunnel was bored (1988 – 1990) it was not used to do specific studies on rock bursting phenomena. There are no doubts that nowadays these specific studies would be done for this deep tunnel (remember: 1300 m overburden, hard rock and eight meters in diameter).

A worthwhile exercise can be done applying the tunnel rock features in order to work out the current parameters used to appraise the rock bursting risk.

4.1 *Criterion of elastic distortion energy*

It was settled by Kwasiniewski and all in Poland (1994). These authors rank the rock bursting risk according to the potential energy of elastic distortion (PSE), which value is expressed as:

$$PSE = \frac{\sigma_c^2}{2 * E_s}$$

being
  PSE = Rock Potential Energy of Elastic Distortion.
  σ_c = Rock Axial Compressive Strength.
  E_s = Rock Distortion Modulus.

In accordance with PSE values the rock bursting risk can be assessed as it is indicated in table 1.

This criterion, named Polish criterion too, do not assess the tensile condition of the rock at its geological site, but the rock ability to store distortion energy for developing bursting phenomena. This criterion understands that, if a rock has not the ability for storing energy, never will undergo a burst whatever its tensile condition is.

The rock strength test at laboratory reached a maximum of 250 Mp. No test were carried out to evaluate the distortion Modulus, nonetheless it can be easily assessed at, for this kind of rock, between 30,000 to 60,000 Mp.

Table 1.

Rock potential energy of elastic distortion	Rock bursting risk
$PSE \leq 50\,kJ/m^3$	Very low risk
$100 \geq PSE > 50$	Low risk
$150 \geq PSE > 100$	Moderate risk
$200 \geq PSE > 150$	High risk
$PSE > 200$	Very high risk

Table 2.

σ_c (Mp)	E_{s} (Mp)	PSE	Potential ability
150	30.000	375,00	Very High
150	35.000	321,43	Very High
150	40.000	281,25	Very High
150	45.000	250,00	Very High
150	50.000	225,00	Very High
150	55.000	204,55	Very High
150	60.000	187,50	High
200	30.000	666,67	Very High
200	35.000	571,43	Very High
200	40.000	500,00	Very High
200	45.000	444,44	Very High
200	50.000	400,00	Very High
200	55.000	363,64	Very High
200	60.000	333,33	Very High
250	30.000	1,041,67	Very High
250	35.000	892,86	Very High
250	40.000	781,25	Very High
250	45.000	694,44	Very High
250	50.000	625,00	Very High
250	55.000	568,18	Very High
250	60.000	520,83	Very High

Applying a range of strength values between 150 and 250 Mp we can produce a risk table about the burst potential ability of the grain-diorite (table 2).

As it can be seen the potential energy of elastic distortion is very high, thus the potential ability of the rock to undergoing a rock bursting phenomenon, if the tunnel reaches a high overburden, is also very high.

4.2 Tangential stress criterion

This criterion was settled by Wang (1998) and deems both the rock massif stress condition and the rock mechanical features. This author works out the value T_s, expressed as:

$$T_s = \frac{\sigma_0}{\sigma_c}$$

being:

σ_0 = Rock tangential stress in the excavation periphery.

σ_c = Rock Axial Compressive Strength.

Table 3.

T_s value	Rock bursting risk
<0,3	No risk
$0,5 \geq T_s > 0,3$	Weak burst risk
$0,7 \geq T_s > 0,5$	Strong burst risk
$T_s > 0,7$	Violent burst risk

Table 4

σ_c (Mp)	Overburden (Z meters)	σ_0 (Mp)	T_S
150	900	27,90	0,2
150	1000	31,00	0,2
150	1100	34,10	0,2
150	1200	37,20	0,2
150	1300	40,30	0,3
200	900	27,90	0,1
200	1000	31,00	0,2
200	1100	34,10	0,2
200	1200	37,20	0,2
200	1300	40,30	0,2
250	900	27,90	0,1
250	1000	31,00	0,1
250	1100	34,10	0,1
250	1200	37,20	0,1
250	1300	40,30	0,2

According to Wang, the rock bursting risk is assessed by means of the T_s parameter in accordance with table 3.

The limit $T_s < 0.3$ is the one formerly proposed by Hoek as the rock bursting phenomena boundary.

Newly, we can apply this criterion to evaluate the rock bursting risk at Maule tunnel.

As do before, different σ_c values are going to be used. We are going to assume 3.1 t/m³ as average specific weight for the grain-diorites.

The σ_0 calculation has to take into account the internal rock stress condition inflicted by the former strains over the massif along the geologic history from its generation, but this is not an easy matter. Like first approach, according to Hoek in the first formulation of this criterion, σ_0 can be evaluated as the overburden weight, thus:

$$\sigma_0 = \gamma * z$$

being

γ = Rock specific weight

z = Overburden

Then the stress ratio T_s would be (table 4):

As it can be seen the rock stress ratio, thus calculated, does not reach the minimum threshold to inflict

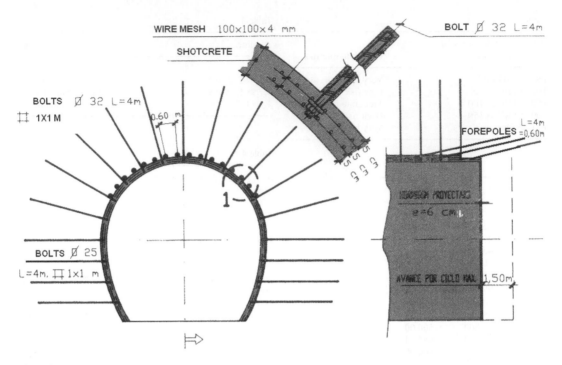

WIRE MESH 100×100×4 mm

SHOTCRETE

BOLT ∅ 32 L=4m

BOLTS ∅ 32 L=4m
1X1 M

0.60 m

L=4m
FOREPOLES =0,60m

HORMIGON PROYECTADO
e=6 CM.

BOLTS ∅ 25

L=4m. 1×1 m

1

AVANCE POR CICLO MAX. 1,50m

Figure 3.

a rock bursting phenomenon, but it happened. In our view it means that σ_0 calculation has left out staple elements and, in this case, can not be only calculated by the simplistic Hoek criterion.

5 ROCK BURSTING PHENOMENA

As saw before the rock massif at Maule tunnel holds a high potential energy of elastic distortion and despite the fact of the deceptive low rock stress ratio, strong rock bursting phenomena took place, along 1,275 meters of tunnel, between the KPs 2,560 and 3,836 tallying with the highest overburdens which are comprised between 900 and 1,300 m.

These phenomena were made up of spontaneous and violent detachments of rock fragments and slabs due to the high stresses to which the rock massif was subjected.

These phenomena had already warned before by the wall popping but, from the quoted KPs onward, they became violent and almost systematic inflicting detachments that occasionally reached a volume of several tens of cubic meters. These detachments were randomly distributed along the vowel, walls and even the bottom of the tunnel, where they derailed muck wagons.

Because the impossibility to foreseeing these phenomena, it was necessary to set up a systematic excavation method in order to minimize the risk and allow to work securely to the people involved in the excavation process.

The excavation method initially chosen was:

– Length of every blast restricted to 3 meters maximum.
– Execution of a forepole umbrella four meters long and spaced 0.6 to 1.2 meters out, according to previous burst signs, before every blast.
– Shotcrete 6 cm thick minimum.
– $\phi 25$ Bolts four meters long; grid 1×1 or 2×2 m according to previous burst signs.
– Wire mesh if necessary.

This method worked right up to KP 3,100 where, tallying with the maximum overburden, and afterwards a huge burst, the last fourteen meters of the tunnel excavation at Maule portal collapsed. As sequel to it $\phi 25$ mm bolts were cut and the 9 mm shotcrete layer broken as well as its wire mesh.

In the wake of that, more stringent measures were taken, which allowed to end the work without troubles. These measures were (figure 3):

– Length of every blast restricted to 1.5 meters maximum.
– Execution of a forepole umbrella four meters long and spaced 0.6 meters out before every blast.

28

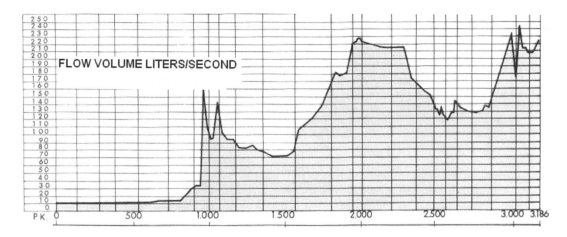

FLOW VOLUME LITERS/SECOND

Figure 4.

- Shotcrete 15 cm thick minimum on the bowel and walls.
- φ32 Bolts four meters long in the tunnel roof; grid 1x1 m.
- φ25 Bolts four meters long in the tunnel walls; grid 1x1 m.
- Double wire mesh into the shotcrete.
- Shotcrete 6 cm thick on the front cut.

In this way the excavation was ended at KP 3,186 without more troubles.

6 ANOTHER EVENTS

6.1 *Water inflows*

According to former reports and in the wake of the flow volumes gauged in the exploration galleries, no important water inflows were expected during the excavation process. As a result the downward excavation process from Maule portal was fitted out to drain a flow volume of 25 l/s. Nonetheless at KP 890 just in the contact andesites – grain-diorites an important inflow sprang out, along with a left wall detachment inflicted by the low angle (10° to 15°) of the longitudinal contact. No porphyry was found between andesites and grain-diorites being the contact above mentioned, in sooth, a fault made up of minor parallel faults with open and weathered joints turning the contact very permeable.

Inflows reached, along the fault, 150 l/s dropping later but never were lower than 70 l/s. Further on, at KP 1,940, appeared a cataclastic area 54 m long with clay seams where the water inflow reached 220 l/s. These inflows were exceeded at KPs 2,990 and 3,040.

In the figure 4 can be seen the inflow chart between the entry portal (Maule KP 0) and the junction point (KP 3,186).

Inflows were important in Melado portal as well, but as it was excavated upward no influence on performances was reported.

6.2 *Massif alterations*

Inside the grain-diorites massif basic plutonic rocks sprang out in sections of 54 and 550 m in length, namely between the KPs 1,942-1,996 and 4,925–5,475 respectively. These rocks have undergone thermal and dynamic metamorphism generating slickenside materials like chlorite, talc, streamer and carbonates.

In these areas could be seen plenty of fault grooves and distortions being these rocks very heterogeneous, some of them hard and others soft. Water inflows were also important in these areas.

Its excavation was problematic. Despite the fact of supporting the excavation with systematic forepoles, bolts four meters long in grid 1.5x1.5 m and several shotcrete layers with wire mesh, it was necessary to built concrete rings with steel ribs and bernold lining.

7 PERFORMANCES

The average excavation performance was 5.13 m/day and portal.

3,186 m were excavated from Maule portal, with an average performance of 4.69 m/day, and 3,539 m from Melado portal, with an average performance of 5,60 m/day.

Assuming ground similarity, the performance difference can be put down mainly to water drainage. In a simplistic manner it can be say that the water drainage inflicted a loss in the excavation performance of 20%.

Along the 1,276 m of tunnel with rock bursting phenomena the average performance was 4.02 m/day and portal. If those meters are deducted, the tunnel average for the rest, 5,479 m, would have been 5,48 m/day and portal. In the same way than before for the water drainage it can be concluded that the rock bursting inflicted a loss in the excavation performance of 36% around.

Ground reaction curves of tunnels considering post-peak rock mass properties

L.R. Alejano & E. Alonso
Natural Resources & Environmental Engineering Department, University of Vigo, Vigo, Spain

G. Fdez.-Manín
Applied Mathematics II Department, University of Vigo, Vigo, Spain

ABSTRACT: In this paper we present a case study in which a tunnel is selected and all of their significant parameters are estimated. Then ground reaction curves for the tunnel are obtained for increasing levels of model complexity, starting from the elastic-perfectly plastic approach, following with a brittle model, and then by means of strain softening (strength weakening) models in which post failure parameters have been calculated according to new techniques. Strength weakening models include constant drop modulus ones, increasing drop modulus with confinement ones and finally a model including also confining stress and plastic strain dependent dilatancy. The effects of the standard support and reinforcement are assessed.

1   INTRODUCTION

1.1   *State of the art in ground reaction curves*

A significant component of tunnels costs stems from support and reinforcement in the form of rock bolts, shotcrete, steel arches, forepole umbrellas and face reinforcement. Should any of this support approach or reach failure or should the tunnel face collapse, then the tunnel and support need to be rehabilitated. This often increases costs much more than the initial support installation, due not just to the cost of the support, but loss of access, down-time of that area, injuries, and possible lost revenues.

Most tunnel designs rely not only on rock mass classifications, but on analytical techniques such as ground reaction curves (GRC) and also on numerical models. The GRC describes the relationship between the decreasing of inner pressure and the increasing of radial displacement of tunnel wall, and it is generally evaluated by theoretical methods, such as analytical or semi-analytical elasto-plastic analyses.

According to Guan et al. (2007), these methods can be divided into two categories according to their treatment for plastic strain. One is the simplified method in terms of total plastic strain, and is represented by Brown et al. (1983) and others. The other is the rigorous method in terms of incremental plastic strain, and is represented by Carranza-Torres & Fairhurst (1999), Alonso et al. (2003) and others. Guan

and co-workers conclude that there is a discrepancy between them in depicting the displacement distribution of plastic region. They also indicated that the rigorous semi-analytical method reflects the nature of tunnel excavation more realistically. Therefore, the use of the so-called rigorous methods is highly convenient if one wants to reliably represent tunnel behaviour.

The GRC techniques often use elastic perfectly plastic models in practice. If failure is allowed to occur, these simple models do not represent well the actual stress strain behaviour of the rock mass, except for bad quality rock masses. In all other cases, strain softening or brittle models are convenient to simulate ground behaviour correctly. The challenge with these nonlinear models is that they require the input of generally unknown material properties.

Recently, some researches are considering this important topic to deepen our knowledge on rock mass stress-strain behavior. For instance, Cai et al. (2007) have proposed to extend the *GSI* system for the estimation of rock mass residual strength. They adjust the peak *GSI* to the residual *GSI_r* value based on two major controlling factors in the *GSI* system. This method for the estimation of rock mass residual strength has been validated by using in-situ block shear test data from cavern construction sites.

The authors of this paper have been working for the last years in the development of techniques which are able to obtain GRC for tunnels excavated in

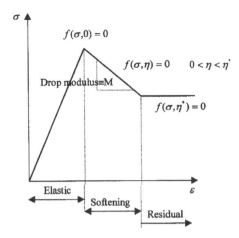

Figure 1. Stress-strain curve of an unconfined test performed on a sample of a strain-softening material.

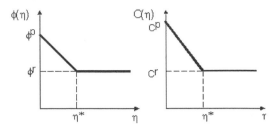

Figure 2. Cohesion and friction ange functions of plastic parameter.

strain-softening continua, as well as in the post failure dilatant behaviour of rock masses (Alonso et al., 2003; Alejano & Alonso, 2005). In what follows, a standard tunnel to be excavated in an average quality rock mass is selected and all their significant parameters are estimated. Then, GRCs for the tunnel are obtained, for increasing levels of model complexity and, we believe, realism.

2 STRAIN-SOFTENING BEHAVIOR

Strain-softening behavior or strength weakening behavior is founded in the incremental theory of plasticity, developed to model the process of plastic deformation. According to this theory, a material is characterized by a failure criterion f, and a plastic potential, g. One of the main features of the strain-softening behaviour model is that the failure criterion and the plastic potential do not only depend on the stress tensor σ_{ij}, but also on the so-called plastic or softening parameter η. Then, the behaviour model is plastic strain dependent.

The failure criterion is defined:

$$f(\sigma_r, \sigma_\theta, \eta) = 0 \tag{1}$$

The strain-softening behaviour is characterized by a gradual transition from a peak failure criterion to a residual one (Figure 1). This transition is governed by the softening parameter η. In this model, the transition is defined in such a way that the elastic regime exists while the softening parameter is null. The softening regime occurs whenever $0 < \eta < \eta^*$, and the residual state takes place when $\eta > \eta^*$, being defined

η^* as the value of the softening parameter controlling the transition between the softening and residual stages. The slope of the softening stage or drop modulus is denoted by M. If this drop modulus tends to infinity, the perfectly brittle behaviour appears, and if it tends to zero, the perfectly plastic behaviour is obtained. It is clear then, that the perfectly brittle or elastic-brittle-plastic and the perfectly plastic behaviour models are limiting cases of this strain-softening model, which can be considered as the most general case.

If we consider a Mohr-Coulomb yield criterion:

$$f(\sigma_\theta, \sigma_r, \eta) = \sigma_\theta - K_\phi(\eta)\sigma_r - 2C(\eta)\sqrt{K_\phi(\eta)} \tag{2}$$

a plastic potential in the form:

$$g(\sigma_\theta, \sigma_r, \eta) = \sigma_\theta - K_\psi \sigma_r \tag{3}$$

where K_ψ is known as dilation coefficient or dilatancy relationship:

$$K_\psi = \frac{1 + \sin\psi}{1 - \sin\psi} \tag{4}$$

The model can be defined by means of piecewise lineal functions of plastic parameter for cohesion $c(\eta)$ and friction angle $\phi(\eta)$ being ϕ^p and c^p the peak parameters and ϕ^r and c^r the residual ones (Figure 2). The elastic regime is characterized by shear modulus G and Poisson's ratio ν. The plastic parameter usually considered is the plastic shear strain:

$$\eta = \varepsilon_1^p - \varepsilon_3^p = \varepsilon_\theta^p - \varepsilon_r^p = \gamma^p \tag{5}$$

3 DILATANCY MODEL

Constant dilation is an approximation that is clearly not physically correct. However, in rock engineering practice, this assumption was made largely because little was known about how the dilation of a rock mass changes past peak (Cai et al., 2007).

A comprehensive review of the literature and observations in regard to published test results indicated that dilatancy is highly dependent both on the plasticity already experienced by the material and confining stress. In a previous work (Alejano & Alonso, 2005), peak dilatancy and friction angle values recovered from a high quality series of tests were compared to confinement stress to show that peak dilatancy is highly dependent on confinement stress. In this way the following expression was proposed:

$$\psi_{peak} = \frac{\phi}{1 + \log_{10} \sigma_{ci}} \cdot \log_{10} \frac{\sigma_{ci}}{\sigma_3 + 0.1} \qquad (6)$$

where ϕ (°) refers to the peak friction angle which can, moreover, be calculated as the slope of the Hoek-Brown failure criterion. σ_{ci} is the unconfined compressive strength of the intact rock, and σ_3 is the confinement stress.

In order to study dilatancy angle decay in line with plasticity, the first option was to assign an exponential decay function to the K_ψ (dilatancy relationship). The decay goes from the previously estimated peak value to a null value corresponding to no plastic volume increase. This null value is proposed in the light of the fact that a rock cannot dilate infinitely. We thus proposed:

$$K_\psi = 1 + (K_{\psi,peak} - 1) \cdot e^{-\frac{\gamma^p}{\gamma^{p*}}} \qquad (7)$$

where the parameter $\gamma^{p,*}$, or plasticity parameter constant, must be calculated for each type of rock.

4 TUNNEL MODELS

4.1 Rock mass and tunnel features

A 7 m radius tunnel is excavated in a basaltic rock mass. A depth of 450 m is considered for this analysis. Based on laboratory tests, average values of unconfined compressive strength of $\sigma_{ci} = 23$ MPa and $m_i = 10$ have been obtained. The Geological Strength Index, GSI, was estimated in a mean value of 55. An average Barton's Q of 0.7 was also estimated from field data. It is wise to conservatively consider very poor quality blasting and local damage in the surrounding rock mass ($D = 0.8$).

The value of GSI, starting from Cai et al. (2004), can be considered as depending on the description of two factors: rock structure (estimated from block volume or block size) and block surface condition (estimated from the joint condition factor or J_C). The main advantage of this method is that in its extension (Cai et al., 2007) it permits to obtain the residual

Table 1. Geomechanical parameters of the rock mass.

Parameter	Unity	Rock mass
GSI_{peak}		55
Q		0.7
$GSIr$		33
σ_{ci}	MPa	23
γ	kN/m³	26.7
E	GPa	3.837
ν	–	0.25
C_{peak}	MPa	0.744
ϕ_{peak}	°	24.81
$C_{res.}$	MPa	0.397
$\phi_{res.}$	°	15.69

strength. We have obtained a residual value of GSI, for the plastified material equal to 33.

The obtained parametric values of the rock mass and rock lab tests, yields a complete set of rock mass data of the rock mass. This data, presented in table 1, is the base for developing post-failure models.

Based on structural observations, an isotropic stress field is contemplated. In what concerns dilatancy various constant values as recommended by Hoek & Brown (1997), together with the variable dilatancy model previously presented are used.

5 BEHAVIOR MODELS OF INCREASING COMPLEXITY AND REALISM

There will be shown in this section, a series of models of increasing complexity and we believe realism, to model the actual behavior of the tunnel.

Model No. 1, presented in Figure 3 represents the elastic perfectly plastic model. This is the one usually applied in practice, even if it only represents adequately bad quality rock masses.

Model No. 2, shown in Figure 4, presents the brittle behavior model. It includes peak and residual strength criteria, but it does not account for strength weakening realistically. According to Hoek & Brown (1997), this behavior can be reasonably assumed for good and very good quality rock masses. However, recent observations indicate that the behavior of this material is not that simple; and new models are needed to suitably model hard rock mass behavior. We have assumed a constant dilatancy value equal to one fourth of the peak friction angle, as suggested by Hoek & Brown (1997). The parameter η^* as proposed by Alonso et al. (2003) is calculated to fit the brittle behavior:

$$\eta^* = \varepsilon_1^{*,pl} - \varepsilon_3^{*,pl} = \frac{\sigma_1^{peak}(\sigma_3) - \sigma_1^{res.}(\sigma_3)}{E} \left(1 + \frac{K_\psi}{2}\right) \qquad (8)$$

33

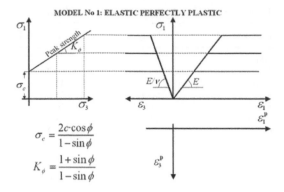

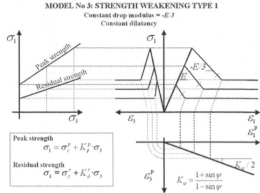

Figure 3. Model No.1, elastic perfectly plastic behavior model.

$$\sigma_c = \frac{2c \cdot \cos\phi}{1 - \sin\phi}$$

$$K_\phi = \frac{1 + \sin\phi}{1 - \sin\phi}$$

Figure 5. Model No.3, strength weakening with constant drop modulus and constant dilatancy.

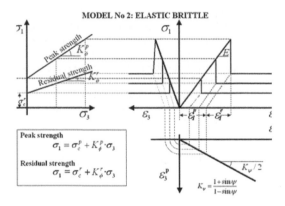

Figure 4. Model No.2, elastic brittle behavior model.

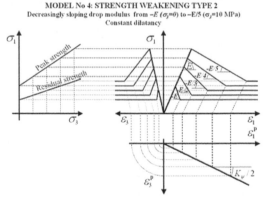

Figure 6. Model No.4, strength weakening with drop modulus decreasing with strength and constant dilatancy.

Model No. 3, illustrated in Figure 5, represents a first simple approach to the so-called strain-softening models. In fact, it is more correct to name this type of models strength-weakening, as suggested by Cai and co-workers, and we have adopted this naming criterion. It has been assumed that the drop modulus is a constant value. We have assumed a drop modulus equal in absolute value to one third of the elastic modulus. We have also assumed a constant dilatancy value equal to one eighth of the peak friction angle, as suggested by Hoek & Brown (1997), for average class rock masses. The parameter η^* (value of the plastic parameter which marks the transition to the residual values of strength) as proposed by Alonso et al. (2003) is calculated to fit the presented strength drop according to:

$$\eta^* = \varepsilon_1^{*,pl} - \varepsilon_3^{*,pl} = \frac{4 \cdot \left[\sigma_1^{peak}(\sigma_3) - \sigma_1^{res.}(\sigma_3) \right]}{E} \cdot \left(1 + \frac{K_\psi}{2} \right) \quad (9)$$

Model No. 4, presented in Figure 6, represents a strength-weakening model more complex than the previous one. Trying to represent the trends observed in large size rock tests the value of the post-failure drop modulus is decreased for increasing values of the confining stress. A simple formulation has been adopted, in which the drop modulus varies in a direct proportional way from $-E$ to $-E/5$, from $\sigma_3 = 0$ to $\sigma_3 = 10$ MPa. A constant dilatancy value equal to one eighth of the peak friction angle has been assumed. As in the previous models, the transition value of the plastic parameter is calculated to fit the presented model, according to:

$$\eta^* = \varepsilon_1^{*,pl} - \varepsilon_3^{*,pl} = \frac{A \cdot \left[\sigma_1^{peak}(\sigma_3) - \sigma_1^{res.}(\sigma_3) \right]}{E} \cdot \left(1 + \frac{K_\psi}{2} \right) \quad (10)$$

Where $A = 0.4 \cdot \sigma_3 + 2$

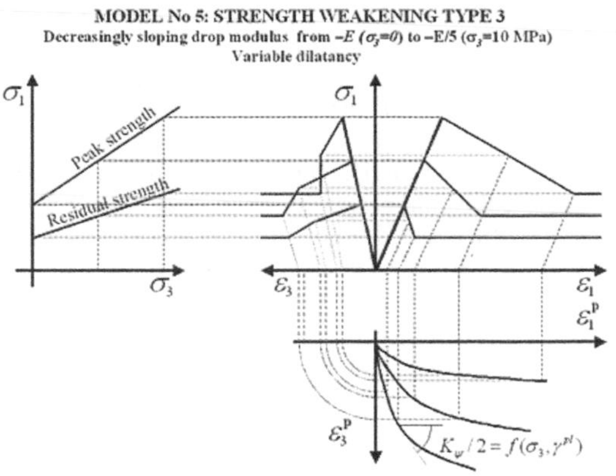

MODEL No 5: STRENGTH WEAKENING TYPE 3
Decreasingly sloping drop modulus from $-E$ ($\sigma_3=0$) to $-E/5$ ($\sigma_3=10$ MPa)
Variable dilatancy

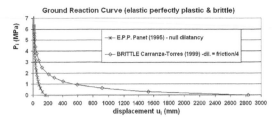

Figure 7. Model No.5, strength weakening with drop modulus decreasing with confining stress and variable dilatancy. This model tries to represent the observed behavior of average quality rock masses.

Ground Reaction Curve (elastic perfectly plastic & brittle)

E.P.P. Panet (1995) - null dilatancy
BRITTLE Carranza-Torres (1999) -dil. = friction/4

Figure 8. Ground reaction curves of the analyzed tunnel for the cases of elastic perfectly plastic and brittle models.

Finally, model No.5, presented in Figure 7, represents a strength-weakening model, seeking to represent actual behavior. The variable dilatancy model as presented in Section 3 has been included. This model suitably represents the actual behavior of average quality rock masses. The dilatancy model is implemented as indicated. A values of $\gamma^{p,*} = 0,02 = 20$ mstrain is assumed for the basaltic rock mass.

The parameters to enter in these models were presented in table 1, with dilatancy values as indicated and plastic parameters to be calculated in each case.

6 GRC FOR THE TUNNEL

6.1 Initial considerations

The ground reaction curves for the indicated tunnel can be calculated rigorously following the proposal by Alonso et al. (2003), according to the different behavior models presented. Little changes have been introduced in the original MATLAB code to account for the particularities of the presented models.

6.2 Analysis of extreme cases

For the first two simple cases there also exist analytical solutions such as those from Panet (1995) and Carranza Torres (1999). The GRCs representing these two models are presented in Figure 8.

It can immediately be observed the enormous differences between the final displacements changing from around 16 cm for the elastic perfectly plastic case (Model No. 1) to around 3 m for the purely brittle case (Model No. 2). This is a clear indication of the dramatic error that can be made when the model selection

is not based on wise criteria. It is also important to put forward that the support and reinforcement effect is highly dependent on the moment of installation, and this is usually controlled by means of the distance to the face and the maximum displacement. This displacement, as we have just seen, may vary one order of magnitude according to the model selection. The actual behavior of the tunnel excavation must obviously be in-between the two extreme cases presented.

If we would obtain now, according to the work by Oreste (2003), the curves corresponding to the support and reinforcement as obtained from the standard Q classification system, it would results non-surprisingly that the recommended support and reinforcement would be able to support the tunnel excavated in the elastic perfectly plastic medium, but not in the brittle one.

6.3 Analysis of the GRCs for strength weakening behavior cases

It has been shown how elastic perfectly plastic and brittle behavior models (extreme cases) does not seem to adequately represent actual average quality rock masses. All the space in between the curves presented in Figure 8 can be filled by means of different strain softening or strength weakening behavior models. This means that the use of this type of model can be considered a wide frame where different conditions can be proposed.

In this paper, the first weakening model used (Model No. 3) is a case with constant drop modulus equal to minus one third of the Young's modulus of the rock mass. The corresponding ground reaction curve shown in Figure 9 is close to the brittle one with a large final displacement.

Model No. 4 represents a new step towards actual behavior for it includes decreasing drop modulus as far as confinement increase as observed in actual rock masses. In this case the GRC (Figure 8) yields a final displacement of 56 cm, which according to our experience seems more in the expectable range for these excavations.

35

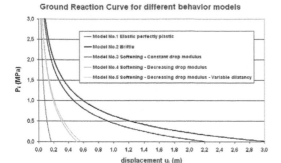

Figure 9. Ground reaction curves of the analyzed tunnel for the five presented behavior models.

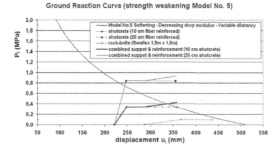

Figure 10. Ground reaction curve of the analyzed tunnel for the more realistic strength weakening model and two possibilities of combined support. Further explanations can be found in the main text.

Model No.5 is an evolution of the previous model in which a variable dilatancy model as presented in section 3 is included. To our knowledge, this model should suitably represent actual rock mass behavior. As it can be seen in Figure 9 the corresponding GRC is closer to that corresponding to Model No.4, even if the final displacement is somewhat smaller, achieving a value 51 cm. From a practical engineering scope these two last results are basically coincident. However, it is important to put forward the fact that the model selection and the correct definition of the parameters representing the strength weakening behavior model utilized is of paramount significance in order to adequately design tunnels according to the convergence-confinement method or to numerical modeling either.

We have obtain now the corresponding support and reinforcement curves, following again the techniques indicated by Oreste (2003), for the case of 10 cm reinforced shotcrete and 1.5 m spaced Swellex rock-bolts, and for the same case with 25 cm of reinforced shotcrete. Results are illustrated in Figure 10.

It is first the first aspect to remark the fact that the proposed support and reinforcement is able to keep the tunnel stability, even if in a somewhat scarcely stability state. This can be due to the fact that the tunnel is quite deep to apply exclusively classification systems.

However, the tunnel, provided that this last GRC was correct would be very close to instability (even if stable) with an equilibrium pressure around 0,4 MPa, and around 10 cm of displacements after the support and reinforcement installation, with a final displacement around 35 cm.

This would be compatible with obtaining a final operative radius around 6.5 meters. According to the Hoek (1999) approach, the safety factor would be scarcely higher than 1. It is obviously not convenient to recommend this design even if it is obvious that a stress-strain equilibrium point is found for the case.

If we define a strain-based safety factor as the ratio between the ultimate strain of the support system and the actual strain of the support system (as proposed by Oreste, 2003) also a very limited value around 1.1 is obtained. Therefore such a support and reinforcement system cannot be recommended.

In fact, for the final design and in the light of this results, it has been recommended to increase the thickness of the shotcrete to 25 mm in order to yield a stress safety factor of 1.3, and a strain safety factor well over 4. All in all with an equilibrium point of $P_i = 0,74$ MPa and $u_i = 246$ mm. These values are compatible with the technical design of the excavation.

7 CONCLUSIONS

In this paper we have shown a case study in which a tunnel is selected and all their significant parameters are estimated. Then ground reaction curves for the tunnel are obtained for increasing levels of model complexity, starting from the elastic-perfectly plastic and the brittle approaches, following with strain softening (strength weakening) models in which post failure parameters have been calculated by means of the newly developed techniques (Cai et al., 2007) and finally, with a strain softening model including confining stress and plastic strain dependent dilatancy. The effects of the standard support and reinforcement are assessed.

This kind of analysis also permits associating the convergence on the wall to the formation of a plastic aureole around a tunnel of radius r under a particular field stress. It is also important to highlight than in all the presented cases the plastification aureole remains constant and therefore, the observed variability in final displacement is exclusively due to the post-failure strain behavior, which turns out to be in this case highly significant in order to control underground excavation behavior.

The main conclusion to put forward regards the high level of error attained in practice engineering when oversimplified models are used to obtain ground reaction curves or to perform numerical models. This is probably the reason why the use of Ground Reaction Curves is still very limited in practice rock engineering.

We finally want to stress the fact, that it is necessary to improve the quality of the estimates of field-scale rock mass post-peak parameters in order to gain confidence in forward modelling and ground support design.

ACKOWLEDGEMENT

The authors thank the Spanish Ministry of Science & Technology, Spain, for financial support of the research project entitled 'Analysis of rock mass post-failure behavior', under contract reference number BIA2006-14244.

REFERENCES

Alejano, L.R., Alonso, E., 2005. Considerations of the dilatancy angle in rocks and rock masses. Int. J. Rock Mech. & Min. Sci. 42(4), 481–507.

Alonso, E., Alejano, L.R., Varas, F., Fdez.Manin, G. & Carranza-Torres. C. 2003. Ground reaction curves for rock masses exhibiting strain-softening behaviour. Int. J. Num. & Anal. Meth. in Geomech, 27: 1153–1185.

Brown, E.T., Bray, J.W., Ladanyi, B., Hoek, E. 1983. Ground response curves for rock tunnels. J. of Geotechnical Engineering 1983; 109(1): 15–39.

Cai, M., Kaiser, P.K., Uno, H., Tasaka, Y., Minamic, M. 2004. Estimation of rock mass deformation modulus and strength of jointed rock masses using the GSI system. Int. J. of Rock Mech. & Min. Sci. 2004; 41(1), 3–19.

Cai, M., Kaiser, P.K., Tasakab, Y., Minamic, M. 2007. Determination of residual strength parameters of jointed rock masses using the GSI system. Int. J. of Rock Mech. & Min. Sci. In press.

Carranza-Torres C. 1999. Self similarity analysis of the elastoplastic response of underground openings in rock and effects of practical variables. Ph. D. Thesis. University of Minnesota 1998.

Carranza-Torres, C., Fairhurst, C., 1999. The elasto-plastic response of underground excavations in rock masses that satisfy the Hoek–Brown failure criterion. Int. J. Rock Mech. Min. Sci. 36 (6), 777–809.

Guan Z., Jiang, Y., Tanabasi, Y, 2007. Ground reaction analyses in conventional tunnelling excavation. Tunnelling and Underground Space Technology 22 (2) 230–237.

Hoek, E., Brown E.T. 1997. Practical estimates of rock mass strength. Int. J. of Rock Mech. Sci. & Geom. Abstr. 34 (8), 1165–1187.

Hoek, E. 1999. Support for very weak rock associated with faults and shear zones. Rock support and reinforcement practice in mining, pp. 19–34. Kalgoorlie, Australia. Ed. Villaescusa, Windsor & Thompson. Ed. Balkema.

Oreste, P. 2003. Analysis of structural interaction in tunnels using the covergence–confinement approach Tunnelling and Underground Space Technology 18 (2003): 347–363.

Panet, M. 1995. Le calcul des tunnels par la méthode des curves convergence-confinement. Presses de l'École Nationale des Ponts et Chaussées. Paris. France.

ROCSCIENCE. RocLab. 2002. Rocscience Inc. Toronto, Canada.

Underground Works under Special Conditions – Romana, Perucho, Olalla (eds)
© 2007 Taylor & Francis Group, London, ISBN 978-0-415-45028-7

Passage of the Cariño fault in the terrestrial emissary "A Malata – EDAR Prioriño Cape"

B. Antuña & I. Pardo de Vera
North of Spain Water Agency, A Coruña, Spain

H. Antuña
Xunta de Galicia, Orense, Spain

J.L. Sánchez
Eptisa, A Coruña, Spain

ABSTRACT: The passage of the faults with TBMs is one of the main challenges for engineering in what respects to the tunnels building due to the great complexity it supposes the passage through them. Thus, this article pretends to collect on the one hand, the previous studies done for the rock mass characterization and on the other hand, to tell the true experience in the Cariño fault passage. This fault, of 190 m long according to the project, is formed by granite materials very heterogeneous and of little consistency. For the execution of the tunnel it is used an open front T.B.M. for hard rock whose excavation diameter is of 3,7 m.

1 INTRODUCTION

During decades, wastewaters of Ferrol has been poured directly into the sea and, as consequence, its estuary is resented and in need of a firm regeneration.

After many attempts, the works, which will convert Ferrol in a drained environment, have begun.

In September 2003, the Xunta de Galicia, Aguas de Galicia and the Confederación Hidrográfica del Norte signed an agreement to finance the improvement of the purification and flow of Ferrol.

Thanks to the investment of the Ministerio de Medio Ambiente and the Xunta of Galicia, with the support of the FEDER and Cohesion Funds from the U. E., waters from Ferrol, Narón and Neda will be depurated before being sent to the sea, with the consequent improvement of the environment of the town of Ferrol and its estuary.The terrestrial emissary A Malata EDAR Prioriño Cape is placed in Ferrol (Spain) and it is part of the infrastructure named "Improvement of the purification and flow of Ferrol".

It consists in building an hydraulic tunnel of 7.344 meter long whose aim is the conduction of sewage from the A Malata pumping station to the sewage works of Prioriño Cape.

In order to build the tunnel, an open front tunnel boring machine (TBM) is used, being the excavation diameter of 3,7 m. and the slope of 1,86‰.

In general, the line of the tunnel goes through granite rocks of a medium/high geotechnical quality, except the Cariño fault which affects about 190 meters of the tunnel.

In the figure 1 we can see the geological profile of the tunnel.

Both the heterogeneity and the little consistency of the Cariño fault materials, such as granite highly fractured or clayey material, make very difficult to cross this area with an open front TBM for hard rock.

The present report collects data about the previous characteristics of the granite massif, obtained by

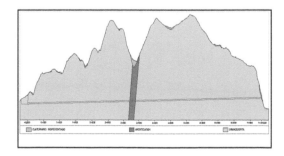

Figure 1. Geological profile.

through geophysics techniques and electrical tomography, completed with a later geotechnical campaign.

Likewise, the incidents occurred in the passage of the Cariño fault are analyzed, emphasizing in the different treatments carried out (addition of a anty-clay agent type Rheosoil 214 and use of bicomponents foams like geofoam) and in the different used supports (forepole bolts, IPN profiles, liner plates, etc).

Lately, it is analyzed the problem arisen in the final meters of the fault as a result of the failure in the joints of the cutting head. To repair this great setback, and given that we are in the fault area, it has been decided to stop the advance with TBM and execute the rest of the fault by manual means, for what it is necessary to execute a parallel gallery to the tunnel, manually.

2 MASS CHARACTERIZATION

Due to the complexity of the fault passage with open front TBMs, we consider indispensable to verify the goodness of the obtained results in the project, and for this, it is used the electric tomography, which is a right method to know the thickness and dip of the different faults and detected fractures. In particular, and with the purpose of studying more carefully the fault placed in the Cariño Valley, it is realized an electric tomography in surface through 715 m (from P.K.3 + 015 to P.K.3 + 730) with a penetration about 100 m.

Figure 2 shows a scheme of the area where the electric tomography was done.

The field data are been interpreted by using the RES2DINV computer programme, which allows the interpretation of inverse resistivity profiles and interactive fittings by square minimums, differences and finite elements.

With the obtained data it is possible to indicate like most probably, the following lithologycal features:

The areas with resistivities superior to $1000\,\Omega m$, must correspond to intact and compact granites.

The sections with resistivities between 500 and $1.000\,\Omega m$, correspond to granites a little bit modified.

The sections with resistivities between 250 and $500\,\Omega m$, correspond to modified granites.

The areas with resistivities less than $250\,\Omega m$, must correspond to IV degree granites, with sandy clays levels, that might correspond to mylonite zones.

In the same way, considering the obtained resistivities distribution, the possible zones of fracture in the studied area are been drawn, being the main ones detected in the P.K.3 + 300 and P.K. 3 + 400, which clearly seem to extend themselves deeply. Between both fractures it appears another zone of lower resistivity which would be a mylonite area (granite of IV to V modification degree). It is convenient to remember that the high humidity in the fault may influence on the results of the electric tomography, producing a more unfavourable interpretation of the soil.

We can conclude that through the study of the electric tomography the fault section that affects the tunnel is approximately 120 m, that is to say, 70 m less than the profile shown in the project, considering the presence of geomechanics low quality soils and because of the TBM characteristics and the impossibility of doing processings from it, it seems convenient the execution of a previous processing to improve the soil stability in the area. There are four possible solutions to solve the passage of the Cariño Valley fault, which will be described next.

– Vertical shaft next to the fault zone, in soil of the best possible quality and opening of an auxiliary gallery parallel to the tunnel from which it will be done the consolidation processing of the fault.
– Vertical shaft next to the fault zone, in soil of the best possible quality and excavation of the fault zone by conventional methods.
– By pass gallery parallel to the tunnel axis from the end of the TBM back up, being the access to the auxiliary gallery, parallel to the tunnel, from it is done the consolidation processing of the fault.
– By pass gallery parallel to the tunnel axis from the end of the TBM back up, being the access to the front, from where it will be done the section of fault by mining methods.

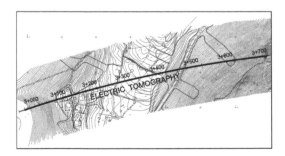

Figure 2.   Situation of the electric tomography.

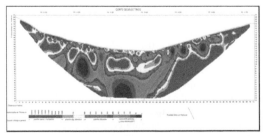

Figure 3.   Electric tomography result.

Although the four proposed solutions are coherent and could be applied with warranties of success, it is also true that there is a series of facts which might affect the choice of the most suitable solution. In addition to the economic aspect, which is the most important one, we can find conditionings such as the work execution time, need of a ventilation shaft, TBM acquisition system (purchase or renting), etc. Next, we will analyze each one:

– Work execution time: if the work execution time is not urgent, we could choose directly the most economic solution, although this one means the TBM stoppage. In case that the execution time is limited, the best option is to combine the TBM work with the execution of the suitable works, which let the TBM pass the fault when it arrives to it.
– Need of a ventilation shaft: if the project doesn't include the existence of a ventilation shaft then it will be inadvisable from an economic point of view, to plan solutions that include the execution of a vertical shaft, except that the cover over the tunnel design is of little importance. In the opposite case, that is to say, if there is thought the presence of a ventilation shaft we must study if there is worthwhile to take advantage of it to pass the fault.
– TBM acquisition system: if the TBM is on renting it will be better profitable with less stoppage times.

At the same time the viability of the four proposed solutions is studied, and to know better the soil behaviour in the fault zone, it is designed a geotechnical campaign based on the execution of four drillings, S1 to S4, with continuous recuperation of witness. Inside those drillings it is foreseen the execution of strain meter and the samples collection for their ulterior analysis in laboratory, in a way that let us do an evaluation the most approximate possible about the geotechnical characteristics of the materials.

The four drillings are taking place in strategic points, according to the data obtained from the electric tomography, that is to say, the S1 reaching the tunnel in the first fault detected on the electric tomography, the S2 between the two faults affecting the area of low resistivity detected by the electric tomography, the S3 in the second fault detected and finally the S4 in an acceptable zone in theory.

In the table 1, it is summarized some of the obtained results in the mentioned drillings.

The analysis of the obtained results is the following: The S1 drilling corresponds to a zone whose meteorization degree goes from intact to fractured, with simple compression strength from 58 to 75 MPa and pressure module from 22.000 to 30.000 kp/cm^2.

The S2 drilling correspond to a zone whose meteorization degree is the same that in the S1 drilling (it presents some modified and oxidized areas), with simple compression strength from 20 to 52 MPa and

pressure module to the tunnel high from 9.300 to 20.000 kp/cm^2, that is to say, with quality a little inferior to the obtained in the S1 drilling, but within a medium high quality.

The S3 drilling corresponds to a zone whose meteorization degree goes from fractured to very fractured, where it was not possible to do the strain meter tests due to the walls did not keep so well like in the other two drillings. The simple compression strength is very low, being the meteorization degree high.

The S4 drilling corresponds to a zone whose meteorization degree goes from intact to fractured, with simple compression strength from 20 to 28 MPa and pressure module a little inferior to the S1 drilling.

In relation to the apparent densities of the rock, the one with the inferior density correspond to the S3, which explains a bigger alteration, the intermediate one corresponds to the S4 and the highest ones are from S1 and S2. In function of the drillings results we can say there are three quality degrees, being the area of the S3 drilling clearly the worst. The S2 and S4 drillings zones could be classified about medium high quality and finally the S1 drilling area seems to be the one with the higher quality.

In figure 4 we can see the soil and tunnel longitudinal profile, as well as the situation of the drillings and the faults deduced from the electric tomography.

Table 1. Obtained results in the drillings.

Drilling	Tunnel material	Meteorization degree	Compression strength	Pressure module
S1	intact fractured	I–II	58–75	30000 22000
S2	fractured oxidized	I–II	20–52	20000 9300
S3	very fractured	II–III–IV	11–16	–
S4	intact fractured	I–II	20–28	23000 20000

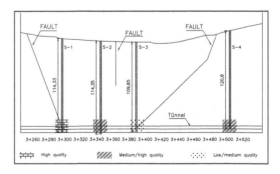

Figure 4. Interpretation of the fault zone.

As we can appreciate in figure 4 the S3 drilling, of worse quality, coincides with one of the faults previously deduced by the tomography; that area may be about 30 m long. However, the other foreseen fault in the tomography coincides with the S1 drilling, with shows a high quality, what might reduce the problematic zone to about 10 m at the most. The most interesting thing is that the shadow central zone between the faults, which was of worse quality in the tomography, now it is represented by the S2 drilling with a medium high quality, similar to the S4 drilling which is outside the fault zone.

So, from the previous analysis we see that the problem of this zone does not seem so important as it was foreseen in the project, since the fault would reduce to about 40 m long, which can be gone through with the TBM without the need of adopting none of the four solutions previously exposed.

3 ANALYSIS OF WHAT HAS HAPPENED IN THE CARIÑO FAULT PASSAGE

As we can see in figure 5, the Cariño fault is divided into two zones of 33 and 32 m respectively. In the first one, material geotechnical characteristics are acceptable, but the second zone presents a great fracturation that together with the water flow make unposible the passage of the TBM, being necessary a consolidation of the soil by resins injection and anty- clay agents.

Between those areas we find a granite material of type II. The table 2 shows outputs reached in the different zones of the Cariño fault.

3.1 Section of the P.K. 3 + 233 to P.K. 3 + 278

In the great majority of this zone (33 of the 45 meters), the placed support has been type IV (liner plates), that is to say, we have got a material with a light fracturation without becoming an important overbreak. Obtained outputs in this section are next to 6 m/day. In the remaining 12 meters it wasn't necessary to put

any kind of support. The figure 6 shows the aspect that presents the rock mass in that section.

3.2 Section of the P.K. 3 + 278 to P.K. 3 + 440

In this section the support is executed by occasional rock bolts obtained advances, about 15 m/day sows the good quality of the rock mass.

3.3 Section of the P.K. 3 + 440 to P.K. 3 + 472

The second zone of geological instability begins in the P.K. 3 + 440, having a development of 32 meters, although it is true there are 8 meters of more intact material. Next, we explain the different types of materials crossed over in this section, as soon as the adopted solutions in every case.

3.3.1 Higly fractured material between the P.K. 3 + 440 and the P.K. 3 + 451

It is characterized by the presence of a contact very similar to the one found in P.K. 2 + 740, that is to say, contact between the granite and the porphyry dikes, although in this case there is no water. So, we encounter a highly fractured material which incessantly falls over the inside of the tunnel, as we can see in figure 7.

In this section it is necessary to place "liner plates" to the advances always helped with rock bolts umbrellas or IPN profiles.

Table 2. Obtanined results in Cariño fault.

P.K.	Meters/day	Support
3 + 233 al 3 + 278	5,5	Liner plates
3 + 278 al 3 + 440	15,0	Ocasional rock bolts
3 + 440 al 3 + 472	0,6	Liner plates, foam and anty-clay agents

Figure 6. Rock mass in the section P.K. 3 + 233 – P.K. 3 + 340.

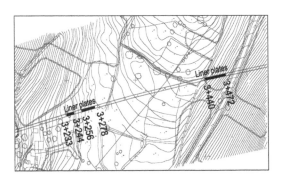

Figure 5. Placement of heavy support on the fault.

In addition to this, if due to the rock mass decompression in the tunnel vault, it is necessary to prick the material over the shield to be able to place the liner plates, we can say we are in a soil where advances are very slow.

3.3.2 *Clay layer of 3 meters between P.K. 3 + 451 and P.K. 3 + 454*

In the P.K. 3 + 451 there is a clay layer of 3 meters thick which makes extremely difficult the excavation.

For a hard rock TBM as the one used in this tunnel, it is very difficult to go through layers of these characteristics, on the one hand, the clayey material sticks enormously on the muck buckets blocking them up, on the other hand it produces the saturation of material on the conveyor belt, being very usual the advance stoppage to clean both the muck buckets and the belt.

Because of the mentioned difficulties to go through the clay layer, it is decided to inject anty-clay agents named Rheosoil 214, frequently employed in EPB TBMs. The Rheosoil 214 is a chemical product that works very well on clayed soils, breaking clays flocculation clot and avoiding the formation of big balls which would block the muck buckets of the TBM

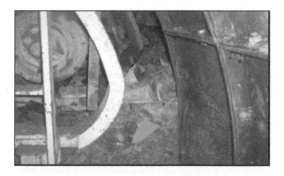

Figure 7.   Material fall inside the tunnel.

head. It is a liquide additive designed for using like a clay dispersant agent, reducing the sticky state of the soil, without reducing its consistence. It can be used directly with the water that is applied to the front through diffusers.

The quantity to be used depends on many factors, like the kind of soil, the speed of the water injection and the desirable consistence of the soil. It generally requires a doses of 0,1 to 5 kg/m^3 of soil. Figures 9 show the clayey material before and afterwards of adding Rheosoil 214.

3.3.3 *Intact material between P.K. 3 + 454 and P.K. 3 + 462*

In this area the material improves the geomecanical characteristics, having no special difficulty the advance of the TBM

3.3.4 *Fractured material of small size between P.K. 3 + 462 and P.K.3 + 472*

This zone is characterized by the presence of a mixture of granodiorite and porphyry dikes of small size, fractured and with great quantity of water and dust material which complicate extremely the excavation.

The high quantity of dust material coming from the overbreak obstructs the advance of the TBM saturating the belt. To minimize such entry of material, both the man passage and the muck buckets are walled up, but the desirable aim isn't achieved.

To the same time of these failed attempts the overbreak continues, acquiring in few days the size shown in figure 10.

Because of this new situation, it was decided a consolidation and impermeability of the affected zone through a foam injection, being this executed by self drilling rock bolts. The foam is an organmineral resin, bicomponente, expansive which applies by means of a two-entries injection pump. The expansion reaction begins in an static mixer, being completed within the soil.

Figure 8.   Rock bolts umbrella to the advance.

Figure 9.   Clay material before and afterwards of adding Rheosoil.

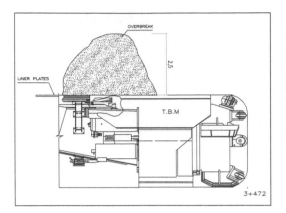

Figure 10.   Final overbreak.

Figure 11.   Injected selfdrilling rock bolts.

Table 3.   Data of the chemical reaction.

Initial temperature	Foam beginning	Foam end	Expansion factor
25°C	35" ± 10"	1'10" ± 15"	15–30

This resin is used for impermeability, soils consolidation and cavities filling, being one of the main advantages of the geofoam its high speed of reaction and its compatibility with water, what allows to consolidate soils immediately ant to stop escapes, reducing to minimum the stoppage time in works, and therefore, the costs.

The resin penetrates through sefdrilling rock bolts working like a cement between them and the wall, penetrating at the same time inside the mass of loose materials accumulated above the shield. The material results consolidated in very few time, and the fissures where the water goes by are closed. The technical data related to the chemical reaction are shown in the table 3.

Once shaken the two components of geofoam are carried in equal parts (volume proportion of 1:1) through a two components injection. Few times before coming into the drilling the components are intensively missed by means of static mixer and injected in the cavity through a nozzle.

Once the reaction is initiated, in few seconds it is formed a foam due to the carbon dioxide and the water steam liberated, not influencing the presence of water, because this resin expands easily into water.

To do the soil consolidation, 8.000 kg of foam were injected through 15 selfdrilling rock bolts, whose length varies between 1 and 6 meters.

The experiences show that the overbreak filling with this foam is a very effective method. The great drawback of the foam in this kind of TBMs, of hard rock with disc cutters, is that the injection in the front may produce important damages to the machine due to the elastic behaviour of this product, which makes the cutters of this TBM unsuitable for its excavation. To improve its drilling, it would be necessary to place some tools, like pears and scrapers.

In the final meters of the fault, it took place the breaking of the cutting head retainers, whose mission is to allow the entrance of lubrication oil to the cavity where the main tread is hosted, as well as avoiding the contamination of the lubrication oil by entrance of water and material coming from outside of the circuit.

Because of the complexity of this kind on breakdown and the consecuences sprung from the TBM advance in this situation (damage of the cutting head main tread), it was necessary to stop the advance and repair the retainers. For this purpose, a by pass gallery was built, which allowed both going through the fault zone by manual methods and driving the TBM to a safe place in which it is possible to anchor the cutting head of the TBM to replace the retainers.

In view of the geomecanical characteristics of the soil, granite material of medium hardness, it was decided to build the by pass gallery of the tunnel with manual methods, that is to say, use of pneumatic drill. The gallery size must be those which let the personnel work in suitable conditions, considering suitable in our case an excavation section of 1,6 m high and 1,5 m wide. Slag material from the excavation is poured on the transporting belt of the TBM.

The by pass gallery support is constituted by TH 16,5 steel arches separate 0,5 m, joined together all the sectors by means of screwed plates. The exterior part of the steel arches is covered with Bernold sheets and wood. In soils with a high alteration level it is convenient before the excavation to execute a light umbrella to the advance, which is formed by 32 mm diameter steel and 1,4 meters long, injected with cement mortar, being the distance between drills of 0,1 m. In figure 12 we can see a scheme of the layout in which it appears the TBM the by pass gallery and the fault plot executed manually in front of the TBM.

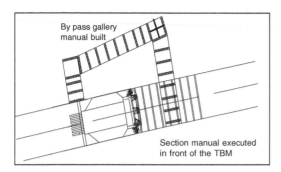

Figure 12. By pass gallery and manually executed section in front of the TBM.

Figure 13. By pass gallery.

Definitely, the gallery excavation process has got the following steps:

a) Execution of a 32 mm diameter umbrella and 1.4 m long.
b) Excavation of the way, 60 or 70 cm.
c) To place support steel arches.

d) To place the filling in the exterior part of the steel arches.

We must also consider that the by pass gallery precises the following services: electric energy water, compressed air, paste injection hosepipes, projected concrete and ventilation.

Once the by pass gallery is finished, it will continue manually the excavation of the tunnel fault zone in front of the TBM with a diameter of 4,1 m (necessary to make the TBM go through). The support in this section of the tunnel, manually built, is formed by TH 16,5 steel arches, separate 0,75 m and 10 cm of projected concrete.

Finally, the TBM is carried to an intact area where it is possible to anchor the cutting head of the TBM to separate it from the rest of the machine and be able to change the retainers. Definitely, 22 days were emploied in the execution of the by pass gallery, 30 days to excavate manually five meters of fault in front of the TBM and 6 days to replace the cutting head retainers.

4 CONCLUSION

As it was mentioned at the beginning of this article, the passage of faults creates a lot of problems. Thus, it is important to do an exhaustive study of the soil to go through as soon as possible. Doing this, we can analyze all the options with their respective consequences, making possible a planification and a consequent action in every moment.

ACKNOWLEDGEMENT

The authors wish acknowledge of translation work of Marisa Lopez Morales.

Underground Works under Special Conditions – Romana, Perucho, Olalla (eds)
© 2007 Taylor & Francis Group, London, ISBN ISBN 978-0-415-45028-7

Design and construction of the Telleda Tunnel

A. Fernández
Works Manager, ADIF

J. Piquer
Head of Works, COPCISA

B. Celada
Tunnel Adviser, AYESA – GEOCONTROL

E. Roig
Head of Tunnel Control, AYESA – GEOCONTROL

ABSTRACT: The Telleda Tunnel forms part of the high-speed railway line between Barcelona and the French Frontier, precisely in the section between Sant Celoni and Riells. It has a length of 214 m and an air section of 85 m^2.

The uniqueness of this tunnel lies in that it had to be built on a hillside, with completely asymmetric stress conditions in the ground. It was to be dug in soft rock and have very little cover. For the purpose of reducing the environmental impact of the work, a mixed method of construction was planned: combining in cut and cover and underground excavation.

The open cut excavation required a provisional slope to be created during the work, with a maximum height of 32 m and a gradient of 1(H):1.5(V), equivalent to 56.3°, which required a very precise soil geomechanical characterization and very accurate numeric modeling to assure its stability during the work.

In this paper a description is given of the soil characterization process, the numeric modeling employed and the behavior of the bank during the different phases of the work, which was excellent.

1 INTRODUCTION

The Telleda Tunnel is located in Sant Celoni (Barcelona) and is part of the Sant Celoni – Riells sub-section of the new high-speed railway line between Barcelona and the French border; designed so that passenger trains can travel at up to 300 km/h.

This tunnel has a length of 209 m, but its uniqueness lies in its scant cover over the vault, which is at most 28 m above the tunnel axis, and the steep transversal gradient of the land.

The combination of these two circumstances converted the selection of the method for construction of this tunnel into a very delicate matter. If an open cut was considered, then the height of one of the side slopes would easily surpass 60 m, with the consequent problems of stability as the excavation would be in soft ground of the Miocene Period, and with an important occupation on the exterior surface which has a high landscape value.

On the other hand, underground construction of this tunnel posed serious problems due to the severe stress dissymmetry of the ground to be excavated and the substantial excavation cross-section, since the tunnel was to have a free section of 87.2 m^2.

This paper presents the activities carried out to define the method for constructing the Telleda Tunnel and the construction thereof.

The Design and Control of the construction of the Sant Celoni to Riells sub-section was commissioned by the Administrador de Infraestructuras Ferroviarias (ADIF) to a Consortium formed by the engineering firms AYESA and GEOCONTROL; whilst the construction of this sub-section was awarded to the Consortium formed by COPCISA and AZVI.

2 DESIGN OF THE TELLEDA TUNNEL

The procedure followed for the Telleda Tunnel project was that of Active Structural Design (A.S.D.), which

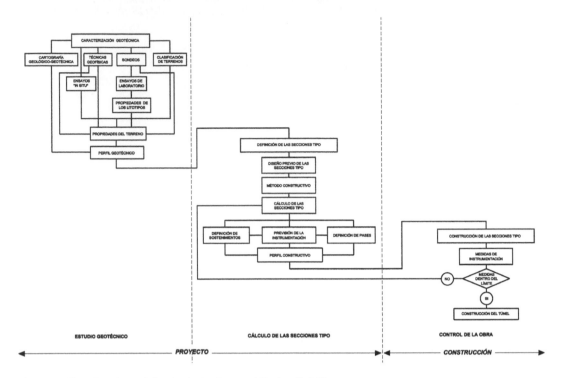

Figure 1. Chart flow for applying the Active Structural Design (A.S.D).

complements the deterministic method used in the definition of the typical cross-sections with the observational method during the construction of the tunnel.

This combination of different methods in the project and work control phase is highly convenient for minimizing the risk involved in the site characterization, especially in the case of tunnels built in soft rock.

The design in the planning phase is based on characterizing the ground with the greatest possible precision and in dimensioning the supports used in all the constructive phases by means of strain-stress calculations that allow convergence that can be measured in each phase of the construction to be defined accurately.

During the work, besides the geomechanical site characterization in each excavation pass to confirm the type of section to apply, if the convergence measured is substantially bigger than the calculated value for each Type Section, it is necessary to make new calculations to define the strengthening of the support.

The A.S.D procedure, as is illustrated in Figure 1, includes the phases of GROUND CHARACTERIZATION, CALCULATION OF THE TYPE SECTIONS and CONFIRMATION DURING CONSTRUCTION, in such a way that construction risks can be effectively minimized.

The most representative characteristics in the Telleda Tunnel project are explained in the following paragraphs.

2.1 Geomechanical site characterization

The Sant Celoni to Riells sub-section was to be built in a terrain of the Miocene Period constituted by a compact granular matrix, integrated by levels of clays, silts and sands which include gravels and granitic boulders.

The effective geomechanical characterization of this kind of terrain is a very delicate task since, on one hand, the granular matrix is quickly degraded by the action of the water which is habitually used in the boring surveys and, on the other, the presence of gravels and boulders makes it practically impossible to obtain representative samples of the terrain for testing in the laboratory by conventional procedures.

For this reason, in this case, it was considered that the most appropriate methods for characterizing this type of material were "in situ" tests, particularly the pressuremeter tests.

Also, back analyses of the stability of various natural slopes existing in the area where the Telleda Tunnel was to be built proved highly useful.

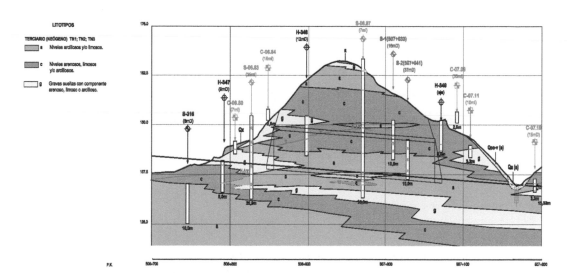

Figure 2. Geological profile of the Telleda Tunnel.

Table 1. Mechanical characteristics of the terrains.

Material	Specific weight (MN/m^3)	Q_U (MPa)	E (MPa)	ν	C (MPa)	$\Phi(°)$
Clays and silts	21.8	0.49	213.9	0.35	0.07	23
Clays and silts	21.6	0.27	268.5	0.30	0.06	34
Gravels	21.4	0.09	220.9	0.30	0.02	40

As a result of the survey work carried out, the geological profile of the tunnel was established, which is shown in Figure 2. In it, the typical materials that appear are basically clays, silts, sands and gravels with little cohesion.

In Table 1 are shown the mechanical characteristics that are considered representative of the stress-strain behavior of these terrains.

The terrain has practically the characteristics of compact soils; although its cohesion and, above all, its modulus of deformation determined by means of pressuremeter tests is quite high.

In accordance with the geological profile presented in Figure 2, it was foreseen that the upper half of the section of the Telleda Tunnel would be excavated in sandy and silty levels, while in the lower half of the section, the excavation would be carried out in clayey and silty layers.

Consequently with the mechanical characteristics determined for these materials, taking into account the overburden over the tunnel, the normal calculations revealed appreciable problems with instability in the heading if the construction of the tunnel was carried out by conventional underground methods.

2.2 Construction process

The election of the construction process for the Telleda Tunnel, in accordance with the foregoing, had to guarantee two specific objectives:

1 Minimize the open cut excavation.
2 Assure the face stability, if the excavation was to do in underground.

The open cut construction of this tunnel required provisional slopes to be dug, the height of which was nearly 60 m, which led to this construction procedure being discarded due to the high environmental impact involved.

Clearly the use of tunnel-boring machines was not feasible due to the short length of the tunnel and for this reason the possibility was explored of using underground construction methods that allowed good control of the stability of the heading. In short, the application of the Traditional Madrid Method was studied, the ADECO-RS method, carrying out the excavation in ADVANCE and BENCHING, under a protective umbrella of jet-grouting and the method of subdivision of the section into several galleries.

The Traditional Madrid Method was discarded being a method requiring much skill and the intensive employment of very specialized manpower difficult to find in the area where the Telleda Tunnel was to be built.

The ADECO-RS and ADVANCE and BENCHING methods, under the protection of a heavy umbrella,

49

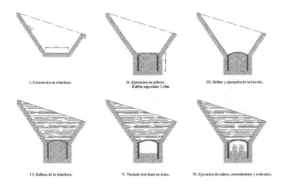

Figure 3. Construction process selected for the Telleda Tunnel.

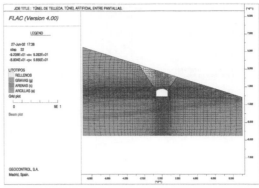

Figure 4. Geomechanical model.

resolved the problem of face instability well, but the construction costs were very high.

Finally, the method of subdividing the section into several galleries was rejected because the severe stress dissymmetry existing in the terrain made it very problematic to guarantee the stability of the temporary partitions for separation between the galleries into which the section had to be divided.

Since the methods studied did not offer sufficient guarantee of a safe and cost-effective construction of this tunnel, a mixed cut and cover/underground solution was considered that was constituted by three main phases:

1 Open-cut excavation to the level of the tunnel vault.
2 Construction of the tunnel walls by means of individual pile walls and the vault of the tunnel, concreting against the excavated ground.
3 Excavation underground of the tunnel section, under the protection of the pile walls and the vault concreted "*in situ*".

With this solution, which is illustrated in Figure 3, it was possible to minimize the environmental impact of the tunnel construction and proceed with the underground excavation in total safety sheltered by the pile walls and the vault of reinforced concrete.

2.3 *Dimensioning the slopes and the tunnel structure*

For the proposed method to be competitive it was essential for the slopes of the open cut excavation to have the greatest possible dip and the structure of the tunnel, formed by the pile wall and the vault of reinforced concrete, could properly withstand the stress dissymmetry they were subjected to. In the following sections the most pertinent aspects are presented regarding the dimensioning of the lateral slopes and the tunnel structure.

2.3.1 *Dimensioning the excavation tunnels*

The design of the excavation slopes for the construction of the Telleda Tunnel was planned leaving a working platform of 16 m in width, and giving priority to the minimization of the left slope which was the highest, being unnecessary to use any means of support in its stabilization.

To study the stability of the left side slope, a parametrized finite element model was prepared for the purpose of studying geometries with slope heights of 36, 34, 32 and 30 m, in which were included the different materials that constituted the ground: gravels, sands and clays. Also included in the model was the structure of the tunnel and the backfill of the excavation, the object being to be able to analyze all construction phases. In Figure 4 is shown the model prepared for the calculations, which were carried out with the FLAC 4.00 program, using the mechanical properties indicated in Table 1 and analyzing the following constructive phases.

1 Excavation of the trench to the level of the tunnel vault leaving an esplanade of 16 m in width.
2 Construction of the sides by means of two pile walls and the tunnel vault, concreted against the ground.
3 Excavation of the tunnel in underground.

As the criterion for selecting the most appropriate slope geometry it was considered that the calculated stability coefficient should be as near as possible to 1.3. This value was considered reasonable because it concerned a provisional slope and whose hypothetical instability would not affect people or buildings.

After carrying out the calculations, it was verified that if the left slope was dug with a slope of 1 (H): 1.5 (V), equivalent to 56.3°, the safety coefficient obtained was 1.34, and so this geometry was accepted.

In Figure 5 is shown the distribution of the shear strains, calculated for the selected geometry. It can be

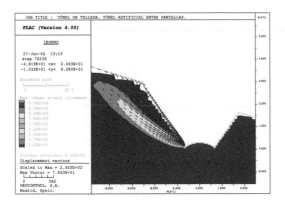

Figure 5. Distribution of the shear strain calculated for the geometry selected.

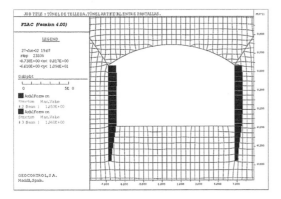

Figure 6a. Axial forces.

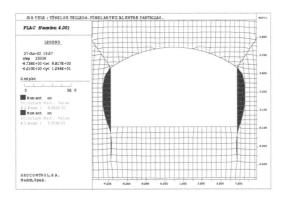

Figure 6b. Bending moments.

appreciated how the maximum values are grouped following the hypothetical rupture surface that would materialize in the event of the slope becoming unstable.

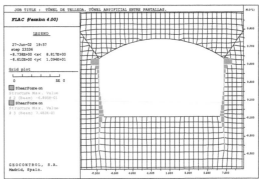

Figure 6c. Shear forces.

2.3.2 *Dimensioning the structure of the tunnel*

Once the construction solution had been chosen, the dimensions of the walls, vault and floor of the tunnel were obtained by calculating the strain-stress conditions for the final phase of construction using the geomechanical model described in section 2.3.1., which were also solved with the FLAC 4.00 code.

In Figures 6a to 6c are shown the distributions of the axial forces, bending moments and shear forces that are acting on the vault and the walls of the tunnel, in the final phase of construction.

The reinforcement was calculated for the walls and vault with the maximum values obtained in the calculations and applying the criteria of the spanish standard EHE-98.

3 CONSTRUCTION OF THE TELLEDA TUNNEL

In the following sections the most representative details in the construction of the Telleda Tunnel are presented.

3.1 *Excavation of the trench*

The excavation of the trench necessary to build the Telleda Tunnel began in September 2004 with the construction of the perimeter gutter to prevent runoff water entering the excavation.

The excavation works were carried out by normal mechanical means with no problems of any type appearing and were concluded at the vault level by mid-November 2004. This signified that the $12600\,m^3$ corresponding to the trench were removed in two months, at a rate of 2500–$3000\,m^3$/day.

A general view of the trench is shown in Figure 7, in the final phase of the excavation, while Figure 8 shows a detail of the highest slope of the trench.

Figure 7. Trench excavated for building the Telleda Tunnel.

Figure 9. Construction of the pile walls.

Figure 8. Detail of the highest slope of the Telleda Tunnel trench.

Figure 10. Construction of the reinforced vault.

the use of a trepan was required due to the need to traverse very resistant boulders.

The construction of the pile walls began in December 2004 and was completed in April 2005.

After construction of the pile walls excavation was started for building the vault, as can be seen in Figure 10.

The construction of the vault began in the month of May 2005 and was completed in June of the same year.

3.2 Construction of the pile walls and the vault

Once the excavation of the trench reached the upper level of the vault, was started the construction of the walls for the tunnel sides, constituted by reinforced piles of 0.85 m in diameter with a spacing of 1.1 m between centers.

Figure 9 shows a moment in the construction of these piles when, in the deepest part of many of them,

3.3 Excavation underground

Before beginning the underground excavation the construction began on the Barcelona portal side, a task which signified digging the portal slopes to the definitive excavation, as can be seen in Figure 11.

The underground excavation began in September 2005, using a conventional backhoe to load the waste onto trucks; as can be seen in Figure 12.

Figure 11. Construction of the Barcelona portal side of the Telleda Tunnel.

Figure 12. Excavation underground in the Telleda Tunnel.

This operation, which was performed without difficulty due to the excellent behavior of the pile walls, was completed in 48 days, which signified an average rate of excavation of 4.4 m of tunnel per working day.

The conclusion of the excavation underground was made to coincide in time with the end of construction of the portal on the French side, as can be seen in Figure 13.

3.4 Finishing the tunnel

Once the portals were built and the construction of the tunnel concluded, the waterproofing of the walls

Figure 13. Construction of the portal on the French side.

Figure 14. Concreting the walls lining the waterproofing sheet.

was carried out using PVC sheeting, and the concreting of the definitive walls which constituted the lining of the tunnel, as can be seen in Figure 14.

The construction of the tunnel concluded with laying the floor that was to allow the emplacement of the tracks, as illustrated in Figure 15.

4 MONITORING THE STABILITY

The critical point in the construction of the Telleda Tunnel was constituted by the stability of the left lateral slope, which had to be excavated with a 56.3° and have a maximum height of 32 m.

To monitor the stability of this slope during the excavation process, benchmarks were placed on the crest which were leveled periodically.

In Figure 16 the evolution is shown of the movements of the six topographical benchmarks located in

Figure 15.   View of the Telleda Tunnel floor.

Starting from the maximum settlements, the following level readings indicated the stabilization of the movements and even the recovery of the movement produced in the benchmarks located in the highest part in the bank.

During the excavation underground, no movements took place in the trench banks, as was foreseen from the calculations carried out.

5   CONCLUSIONS

The Telleda Tunnel, in spite of having a length of only 214 m, proved highly difficult to design due to the steep of the existing natural slope, transversal to the

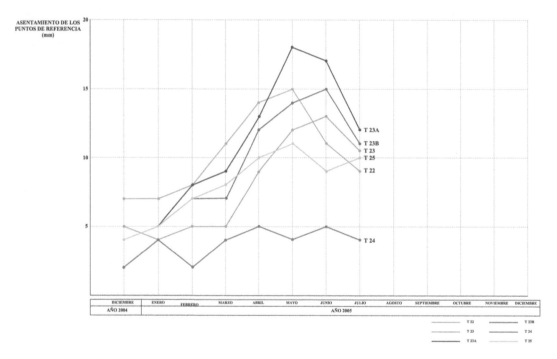

Figure 16.   Evolution of the settlement of the control benchmarks placed on the highest slope.

the highest part in the bank; that was the most critical area in the slope.

In this figure it can be appreciated that as the excavation proceeded the settlement of the control landmarks occurred, reaching a maximum value of 17 mm when the excavation arrived at the devation of the tunnel vault.

This value practically coincided with that foreseen in the calculations, which signified the confirmation of the forecasts made.

tunnel, in which it was to be dug and to the significant ecological value that the external surface covering the tunnel had.

During the preparation of the project several alternative construction procedures were analyzed and finally a mixed method was chosen, cut and cover and underground, which permitted the external ground surface affected by the construction to be minimized.

The ground to be excavated was constituted by a compact granular matrix, made up of layers of clays, silts and sands which included gravels and granitic boulders.

This terrain, which was considered to be a compact soil or a very soft rock, presented small, but significant cohesions, which allowed excavation slopes to be planned with gradients of 56.3° and a maximum height of 32 m.

Once the open cut excavation had been carried out the pile walls were built which would later constitute the tunnel sides, and the tunnel vault was accomplished.

After building the pile walls and the vault, the excavation of the tunnel started underground, using conventional machinery, to complete the tunnel.

The construction method planned proved highly satisfactory as no geotechnical problems arose during the work, which, together with a careful executed construction, allowed this tunnel to be built in a very safe and efficient manner.

Safety in tunneling machines

C. Fernández, C. Martínez & J.C. Sancho
UPM-Laboratorio Oficial JM Madariaga, Madrid, Spain

ABSTRACT: Authors of papers to proceedings have to type these in a form suitable for direct photographic reproduction by the publisher. In order to ensure uniform style throughout the volume, all the papers have to be prepared strictly according to the instructions set below. A laser printer should be used to print the text. The publisher will reduce the camera-ready copy to 75% and print it in black only. For the convenience of the authors template files for MS Word 6.0 (and higher) are provided.

1   INTRODUCTION

The free circulation of products in Europe is rapidly developing to the benefit of consumers and of those businesses which know how to adapt themselves to the new resulting situation and, as a consequence, to achieve competitive advantages.

One of the main elements for this inner frontier elimination is the strategy developed to reach technical harmonization with all the laws and regulations related the security of people and products, as defined by the directive policy in 1985 known as "new approach" (Resolution of the EEC Council 85/C136/01), which defines that all harmonization directives shall be based on Article 100 of the European Communities Treaty.

The basis for this above mentioned policy is the reduction of Community legislation, limiting the content of every directive to the definition of essential requirements for the free circulation of products and equipment within the territory of the European Union. At the same time, these essential requirements have the mission of insuring the protection and the security of people, animals and goods in all the territory of the EU.

The Machinery Directive is one of the legal documents of the EU whose objective is to create a path to the single market of the EU, in other words, a market with freedom for the movement of people, products, capitals and services.

As a previous condition, a machine to be put on the market or installed shall de accompanied by a CE conformity declaration, exhibit the CE marking and its instructions manual shall be delivered as a whole with the machine. The Member States can not prohibit, restrict or disallow its trading or installation on its territory of products showing the CE marking except in the case the requirements related to CE marking had been incorrectly applied. The manufactures assumes the exclusive and absolute responsibility of the conformity of the product with the applicable directives. The manufacturer is responsible for the design and manufacturing of the product in compliance with the health and security essential requirements set up by the applicable directives as well as from the execution of the conformity evaluation of the procedures established by those directives.

2   ESSENTIAL HEALTH AND SAFETY REQUIREMENTS

2.1   *General requirements*

A fundamental principle of the New Approach consists of restricting the legislative harmonization to the essential health and safety requirements. These requirements are aimed to offer and guarantee a high degree of protection. They are compulsory and only the products which fit the general requirements can be placed on the market or put into service.

The essential safety and health requirements to consider are those established in Annex I of Directive 98/37/EC. Next, some essential health and safety requirements are shown for those cases which have any variant to general machines. In the following points, requirements are marked which deserve special commentary due to the characteristics of machines mentioned in this article.

The principles of safety integration is applicable to any kind of machine, and the need of indicating to the user the cases in which the machine requires special conditions of use, shall be specially taken into account.

The materials used in fabrication should not be capable of coming in contact with other materials and produce sparks. The use of hydraulic fluids, usual in machine operation, should be taken into account. It is compulsory to use non-flammable fluids or the provisions of other means to provide an equivalent level of safety.

To prevent the risk of crushing, consideration must be given to those cases in which it is expected to pass through the partition to the rear zone of the cutting head and in a similar way through the cutting head to the zone of the face, where proper openings have to be provided for this and also ensure a safety space between the partition and the cutting head. Whenever possible, the cutters must be of the type which permit them to be changed from the rear without having to pass to the front of the cutting head. In all cases, the head must be equipped with a system for preventing its unintentional rotation.

When access is required via the front of the cutting face under conditions of unstable rock, protection measures must be provided such as temporary securing of rocks or pretreatment of the surface of the rock in front of the cutting head.

In order to handle heavy elements, weighing more than 50 kg, handling and control devices need to be installed. These devices have to be stationary with winches and levers for moving the elements, or rotary when the pieces can be lifted directly to any position around the section of the tunnel. In the special case of handling segments, the systems must be designed in such a way that the need for people to be present in the loading zone is avoided.

In the case of tunneling machines with shield, they must be provided with rotation indicators and a system of counter-rotation that will permit the machine and its support train to be returned to the correct orientation. One of the main risks that can occur is the sudden rotation of the machine when the cutting head or the jib remains stuck in the face. For that reason, these machines must have a device which disconnects the drive motor in the event of the machine exceeding the angle of rotation established by the manufacturer. The minimum pressure conditions for anchorage will vary depending on the geological conditions of the ground, so it will not be possible to start up the cutting head or apply the thrust until the minimum anchorage pressure has been reached. In the same way, if, during the functioning of the machine, the pressure were to drop to below the established minimum, the cutting head must be stopped and the thrust on the face must be halted automatically.

Against the risk of falling objects, tunneling machines with a shield must be provided with protective structures against that risk in places where work is permanently being carried out. These structures must guarantee a protection level 1 in accordance with EN 13627:2000 "Falling-object protective structure". In tunneling machines without a shield, the main control post must be provided with a cabin or roof for protection against the risk of falling rocks and objects such as tools and replacement parts. Said roof must be capable of withstanding a point load of 2000 N.

Prevention risk involves the need of using guard and protective devices fixed or movable, and adjustable to restrict access. The fixed guards are generally metallic sheets that prevent access to parts of the machine that should not be accessible during operation. The utilization of special bolts for its fastening is recommendable in order to prevent the unnecessary opening of these guards, or any other means which provide an equivalent safety degree.

The starting control for the machine must be located in the main control post and all the auxiliary controls for the equipment must be subordinate to this control. The control has to be associated with an early warning system in all equipment with rapid movement or which might imply a risk, such as conveyor belts, cutting heads, support train, etc. The starting will be timed with the early warning system, in such a way that starting is delayed by a length of time determined in accordance with the level of risk and the specifications contained in the regulations for the product.

This type of machine must be designed so that it can incorporate adequate equipment for ventilation, dust elimination and collection or elimination systems within the cutting head with the aim of avoiding and preventing its propagation, for example sprayed water and screens.

Internal combustion engines used for underground works are contained in annex IV of the machines directive and therefore, in accordance with the procedure for evaluation of conformity, the intervention of a notified body is required. These engines must be used always provided a minimum flow is guaranteed along with an extraction system away from the tunnel portals.

Underground works can present risks of the emanation of gases from the ground itself, and these gases can be toxic, flammable or they can simply displace oxygen and reduce its concentration in the air. For this reason, tunneling machines must have atmospheric control equipment capable of detecting the presence of these gases. When the risk of explosion due to the presence of flammable gases has not been taken into account in the design of the tunneling machine, then the machine must be fitted with detection sensors for flammable gases which have to be located at the points closest to the work face, in the flow of the dust elimination systems and at the unloading/transfer points of the conveyor belt system. Both the electrical and the mechanical equipment located in the zone of the evacuation airflow for the dust must be provided with any of the means of protection against explosion indicated in standards EN 60079-0 and EN 13463-1 respectively, depending on the category or conformity which the equipment has to have in line with the assessed risk of the presence of an explosive atmosphere. The control system must de-energize all the electrical and mechanical equipment not intended for use in explosive atmospheres or which is intended for working in an explosive atmosphere but which are not category 1 or M1, in the event of reaching the limit values for the concentration of methane established by national regulations.

When the machine has been designed for working in the presence of an explosive atmosphere or an atmosphere of combustible dust, then it must comply with the essential requirements of health and safety established in Directive 94/9/CE "on the approximation of the laws of the member states concerning equipment and protective systems intended for use in potentially explosive atmosphere" and follow the conformity evaluation procedure established therein.

Electrical equipment must comply with standards EN 60204-1 (rated voltage not exceeding 1000 V ac nor 1500 V dc and for rated frequencies not exceeding 200 Hz) and EN 60204-11 (supply voltages above 1000 V ac or 1500 V dc without exceeding 36 kV ac or dc with rated frequencies not exceeding 200 Hz), depending on the voltage used. These standards establish the requirements relating to electrical equipment guaranteeing safety with regard to electrical risks.

Only transformers that are air-cooled, without oil, will be permitted. They must include insulation devices for the power supply on both the high tension and the low tension sides. The insulation of the electric cables installed must be of the type LSF, with low production of smoke and vapors.

Tunneling machines must be fitted with means of fire detection and alarm. In places where the risk is greatest, for example internal combustion engines, main drive motors for the cutting head, hydraulic power units, electrical cabinets and transformers, fire extinction systems or portable extinguishers must be installed. In the case of tunneling machines with shield, a water sprinkling system must be installed capable of producing a curtain of water in the entire section of the tunnel in the rear part of the hauled support train, which must be operated manually. Also, the hydraulic fluids used in them must be fire resistant as per ISO 7745:1989.

A zone must be marked out in the support train where there exists a store for personal safety equipment including first aid equipment, individual protection equipment and breathing apparatus (self-rescuers).

2.2 Requirements for use in explosive atmosphere

Where a tunnel has been or is being driven through ground where methane or other flammable gas is present in significant quantities, all electrical and mechanical equipment used underground should be so constructed, installed, protected and used as to guard against the danger of explosion. Electrical and mechanical equipment should be capable of being safely isolates should the presence of flammable gases make it necessary.

When the main risk is due to the present of firedamp or dust coal, in order to fulfill the requirements established by the explosive atmosphere directive, EN 1710 can be used. This European standard specifies requirements for the constructional features of equipment and components that may be an individual item or form an assembly, to enable them to be used in mines, or parts of mines, susceptible to explosive atmosphere or firedamp and/or combustible coal dust.

All electrical and non-electrical equipment and components for use in potentially explosive atmosphere shall be designed and constructed to good engineering practice and in conformity with requirements of group II category 2, or group I category M2 when the main risk is due to the present of firedamp or dust coal, equipment to ensure that the ignition sources do not occur. In particular, the following requirements described in EN 60079-0 and EN 13463-1, apply to all machines and shall be taken into account:

- The need to restrict the maximum surface temperature.
- The need to meet the electrostatic requirements.
- The need to restrict the use of exposed light metals.
- The need to perform test on non-metallic parts on which the ignition protection depends.

Requirements for fans, for use in explosive atmosphere, are given EN 14986.

Diesel engines used in potentially explosive atmosphere shall be flameproof Group I or Group II internal combustion engines and comply with EN 1834-2 or 1834-1 and 1834-3 respectively.

Where drilling equipment and components are to be used in circumstances where there is a likelihood of there being an ignition risk between the drilling tools and the material being drilled, the manufacturer shall ensure that the drilling machines and drilling tools are not capable of creating hot surface or sparks.

Where possible, fire-resistant materials shall be used. Where this is not possible, the fire protection used shall prevent the atmosphere from igniting the atmosphere and the requirements of EN 13478 shall be complied with. Machines fitted with combustion engines shall be equipped with a device for portable fire extinguishers and, where necessary, an automatic fire extinguishing system. Hydraulic equipment should be designed and constructed to operate with hydraulic fluids for which the fluid manufacture has provided proof that they are fire-resistant, or the provisions of other means to provide an equivalent level of safety.

The conveyor belt shall be designed and constructed in conformity with requirements of EN 14973. This European standard specifies electrical and flammability safety requirements for conveyor belts intended for use in underground installations, in the presence of flammable or non-flammable atmosphere.

The technical committees of standardization bodies are in charge of drawing up the technical standards covering the essential requirements for health and safety set down by the directives. The standards constitute what are known as harmonized standards, once their reference has been published in the OJ (Official Journal), which, though they are not obligatory, nevertheless represent a preferred solution for complying with the said requirements set down by the directives.

In relation to the machines directive, various complementary categories of harmonized standards are established depending on the general or particular aspects covered by them. Within this categorization are the "type C" standards referring to the specific safety requirements for a machine or a group of machines, the application of which implies the presumption of compliance with the essential health and safety requirements set down by the directive. The technical standards on machines for tunnel construction are developed by the CEN/TC 151 "construction equipment and building material machines – safety". Below, reference is made to some of the standards published in relation to tunnel construction machines:

- EN 815:1996. Safety of unshielded tunnel boring machines and rodless shaft boring machines for rock.
- EN 12336: 2005. Tunneling machines. Shield machines, thrust boring machines, auger boring machines, lining erection equipment. Safety requirements.
- EN 12111:2002. Tunneling machines. Road headers, continuous miners and impact rippers. Safety requirements.

3 CONFORMITY ASSESSMENT PROCEDURE

These are the procedures which manufacturers have to follow in order to demonstrate the conformity of the product with regard to the provisions of the directive or directives that are applicable to it, fundamentally compliance with the essential health and safety requirements of the product and quality controls over it when it comes to complying with those requirements.

For these purposes, the machines directive distinguishes two categories of machine, ordinary machines which comprise 90% of them and the machines included in annex IV of the said directive. The latter are regarded as machines presenting special risks and requiring the intervention of a notified body.

As tunneling machines are not included in annex IV of directive 98/37/EC, they do not require the intervention of a notified body. The manufacturer has to apply module A, internal control over production, by means of which he draws up the technical file on construction of the machine, which he retains and places at the disposal of the competent national authorities and adopts all the necessary measures for guaranteeing that the manufacturing process ensures conformity of the products with the technical documentation and with the applicable requirements.

When the machine is intended for use in explosive atmospheres, it will come within the scope of application of the directive on explosive atmospheres (Directive 94/9/EC) and the evaluation procedures for conformity described in that directive for electro-mechanical equipment of category 2 or M2 must be followed. The installed electrical equipment must be evaluated as an individual component of the equipment and it must present the corresponding EC declaration of conformity with regard to the EC type certificate issued by a notified body (for example, motors, lights, electrical switchboards, etc.) and with regard to the notification of the manufacturer by a notified body whose identifying number forms a part of the marking, along with its instruction manuals and its own marking. Nevertheless, this individual certification does not treat the interconnection of the different components of the electro-mechanical assembly as being a unit. With the aim of complying with section 1.6.4 of the essential safety requirements of the directive on explosive atmospheres, the equipment and components, including their interconnections, must be evaluated by the manufacturer from the viewpoint of ignition risk.

Moreover, non-electrical equipment also requires an evaluation of the ignition risk in order to comply with the requirements established by the directive on explosive atmospheres and which have to form part of the technical file for construction of the mechanical assembly. This evaluation ought to be based on the principles of protection towards explosion set down in the series of standards EN 13463 (non-electrical equipment destined for explosive atmospheres). The interconnection of electrical/non-electrical equipment also requires such an evaluation of ignition risk.

Once the technical file has been created and compliance with the essential requirements set down by directive 94/9/EC has been verified, the file must then be sent to a notified body for the explosive atmospheres directive who will issue an acknowledgment of receipt for it and safeguard it.

4 TECHNICAL CONSTRUCTION FILE

The manufacturer is obliged to draw up a technical construction file. It document must demonstrate the conformity of the machine to the applicable requirements. Technical construction file must contain information to demonstrate the conformity of the machine to the applicable requirements. It must contain the following documents:

- An overall drawing of the machinery together with drawings of the control circuits.
- Full detailed drawings, accompanied by any calculation notes, test results, etc., required to check the conformity of the machinery with the essential health and safety requirements.
- A list of the essential requirements of this Directive, standards, and other technical specifications, which were used when the machinery was designed.
- A description of methods adopted to eliminate hazards presented by the machinery and a list of standard used.
- A description of methods adopted to eliminate hazards presented by the machinery.
- If he so desires, any technical report or certificate obtained from a competent body or laboratory.
- A copy of the instructions for the machinery.
- For series manufacture, the internal measures that will be implemented to ensure that the machinery remains in conformity with the provisions of the Directive.

5   EC DECLARATION OF CONFORMITY

The EC declaration of conformity must ensure that the product satisfies the essential requirements of the directive. It must be draw up for the manufacturer or his authorized representative established within the Community (Fig. 1).

The contents of the EC declaration of Conformity are laid down to in Annex II. As a minimum, it must contain the following particulars:

- Name and address of the manufacturer or his authorized representative established in the Community.
- Description of the machinery (make, type, serial number, etc.).
- All relevant provisions complied with by the machinery.
- Where appropriate, a reference to the harmonized standards and the national technical standards and specification used.
- Identification of the person empowered to sign on behalf of the manufacturer or his authorized representative.

This declaration must be draw up in the same language as the original instruction. It must be accompanied by a translation in one of the official languages of the country in which the machinery in to be used.

EC Declaration of Conformity

We (Company, name and address of the manufacturer, other identification)

declare under our sole responsibility that the following product:

_____
(name, type or model, lot, batch or serial number, possibly sources and numbers of items)

Complies with the requirements of the EC machinery Directive 98/37/EC.
(and other applicable directives)

Applicable Harmonized Standards:

Applicable National Technical Standards and Specifications:

Name

Title of signatory

Place and Date of issue

Authorized signature

_____

SIGNATURE

Figure 1.   Model of EC declaration of conformity.

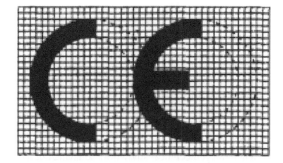

Figure 2.   CE marking.

The signature of the EC declaration of conformity authorizes the manufacturer to affix the CE marking to the machinery.

6   "CE" MARKING

The CE marking (Fig. 2) indicates that the products is conforms (according) to the essential requirements and that the product has been subject to the appropriate conformity assessment procedure. It may not be affixed until the conformity assessment procedure has been completed, and it has been verified that the product complies with all the provisions of the directive or directives.

The CE marking is mandatory and must be affixed before the machinery is placed on the market and put into service. The CE marking shall consist of the initials "CE" taking the following form:

The various components of the CE marking must have substantially the same vertical dimension, which may not be less tan 5 mm. If the CE marking is reduced or enlarged the proportions given in the above drawing must be respected.

All machinery must be marked with the following minimum particulars:

– Name and address of the manufacturer.
– The "CE" marking.
– Designation of series or type.
– Serial number, if any.
– The year of construction.

The machinery must also bear full information relevant to its type and essential to its safe use.

7 SUMMARY AND CONCLUSIONS

Tunneling machines must be designed to comply with the essential health and safety requirements of Directive 98/37/EC and the manufacturer can use one of standards EN 815, EN 12336 and EN 12111 and follow the procedure known as Module A. The manufacturer must provide along with the tunneling machines:

– The CE marking.
– The EC Declaration of conformity with Directive 98/37/EC with reference to the standards used for complying with the essential safety requirements, especially those of Directive 98/37/EC.
– The instructions for installation, maintenance and repair, in Spanish.

When the geotechnical study prior to laying out the tunnel cannot establish the risk of presence of an explosive atmosphere as being insignificant (zero), then the tunneling machine must be designed for working in the presence of a potentially explosive atmosphere, and the manufacturer must furthermore comply with the essential safety requirements of Directive 94/9/EEC, as equipment of conformity category 2 or M2. It can use standard EN 13463 or EN 1710 for this, following the procedure known as Module A. The manufacturer must provide along with the tunneling machine:

– The CE marking.
– The EC Declaration of conformity with Directive 94/9/EEC with reference to the standards used for complying with the essential requirements of Directive 94/9/EEC, and also for complying with the requirements regarding other risks, which are considered in Directive 98/37/EC, in Spanish.
– The Declaration of Conformity must be signed by the manufacturer, who is understood as being whoever is responsible for the design and construction of the tunneling machine. The manufacturer can be advised by expert persons or entities, but he takes on the overall responsibility for the product.

REFERENCES

Directive 98/37/EC of the European Parliament and of the Council, of 22 June 1998, on the approximation of the laws of the Member States relating to machinery. Official Journal of the European Communities. 23.7.98.

Directive 94/9/EC of the European Parliament and the Council, of 23 March 1994, on the approximation of the laws of the Member States concerning equipment and protective systems intended for use in potentially explosive atmosphere.

Guide to the implementation of directives based on new approach and global approach. Brussels, 12 October 1998. Doc. Certif. 98/1. Version 1.0.

Community regulations on machines. Comments on directives 89/392/EEC, 91/368/EEC, 93/44/EEC and 93/68/EEC. Jean-Pierre Van Gheluwe.

EN 815:1996. Safety of unshielded tunnel boring machines and rodless shaft boring machines for rock.

EN 12336:2005. Tunneling machines. Shield machines, thrust boring machines, auger boring machines, lining erection equipment. Safety requirements.

EN 12111:2002. Tunneling machines. Road headers, continuous miners and impact rippers. Safety requirements.

EN 1710:2005. Equipment and components intended for use in potentially explosive atmosphere in underground mines.

EN 13463-1. Non-electrical equipment for potentially explosive atmosphere. Part 1: basic method and requirements

BS 6164:1990. Code of practice for safety in tunneling in the construction industry.

Underground Works under Special Conditions – Romana, Perucho, Olalla (eds)
© 2007 Taylor & Francis Group, London, ISBN 978-0-415-45028-7

Rehabilitation of Lapa tunnel, Metro do Porto

P. Ferreira
Metro do Porto, Porto, Portugal

M.S. Martins
University of Minho, Guimarães, Portugal

L. Ribeiro e Sousa
University of Porto, Porto, Portugal

ABSTRACT: This paper presents a brief description of the main defects and deteriorations in old railway tunnels with particular emphasis with the Portuguese railway tunnels. The case study of Lapa tunnel in the city of Porto was analysed. This tunnel was included in the Porto Metro network. Inspections were performed and they permitted to conclude that almost the totally of the tunnel presents deteriorations in almost all the support, humidity and water infiltrations. The rehabilitation works performed in the tunnel are described as well as numerical studies for the tunnel.

1   INTRODUCTION

Railway tunnels were constructed in Portugal mainly in the 19th and beginning of 20th century. The age of such works implied the degradation of their state and consequently affected their stability conditions, which may cause safety issues in those underground structures. The non-detection of the defects, can affect the stability of the old railway tunnels and the necessity of presenting solutions for their resolution are indispensables tasks.

It is fundamental a systematic inspection of the works as well as their monitoring in order to guarantee adequate safety levels.

The paper presents a brief description of the main defects and deteriorations in old railway tunnels with particular emphasis to the Portuguese ones. The case study of Lapa tunnel in the city of Porto is analysed. This tunnel was included in the Porto Metro network. Inspections were performed and they permitted to conclude that almost the totally of the tunnel presented deteriorations with a thickness of about 5 mm, humidity and water infiltrations. The rehabilitation works for the tunnel are described as well as the performed numerical studies.

2   RAILWAY TUNNELS IN PORTUGAL

The railway network of Portugal started with the construction of the stretch Lisbon-Carregado in 1856.

The first tunnel with a 650 m length was soon completed.

Portuguese railway network has about 120 tunnels with almost all of the tunnels being a hundred years old. Therefore many deteriorations and defects are likely to occur, mainly due to the ageing process and to the construction method used in old tunnels; the Belgian method was the one that prevailed in the excavation of tunnels (Silva, 2001; Silva et al., 2002).

The main defects detected in the Portuguese tunnels are essentially related with the construction method, namely in:

i) Rock masses – they are represented by a certain decompression around the cavity, associated to successive deconsolidation due to the construction method. Significant voids may occur which are likely to have high negative effects on the stability of the underground structures.

ii) Supports – defects in the crown and in the extrados of the support, as well as the occurrence of hollow joints, deficiency in springings, defective drainage and waterproofing. The deterioration of supports with decrease in resistance is mainly associated with the presence of water, the erosive action of the wind caused by the passage of trains, the mechanical actions deriving from the envelope rock mass, significant deformations, cracks, etc.

The observation and the inspection of the tunnels have become essential. Regular inspections are an extremely effective means. They make it possible to carry out a timely detection of defects in these structures and to propose some corrective measures. At the most sensitive and problematic zones, appropriate monitoring using measuring equipment must complement the inspections.

The computerized processing of the data collected in the inspections and the observation of the tunnel, and also the use of artificial intelligence techniques are aiding tools for data storage and also a valuable help for decision-making regarding the safety and the rehabilitation works (Silva et al., 2002; Sousa et al., 2004).

3 PORTO METRO

3.1 Layout

The Porto Metro is light rail system about 70 km long in Porto metropolitan region connecting seven municipalities, one the largest networks built at once in Europe. It has two underground tunnels in the city of Porto, in line C, 2.6 km long and in line S, 3.5 km long (Babendererde et al., 2006).

The tunnels for the lines C and S were excavated by two Herrenknecht EPB TBMs, that have an internal diameter of 7.8 m and accommodate two tracks with trains running in opposite directions. Line C has five underground stations, a maximum cover of 32 m and a minimum of 3 m before reaching Trindade station, while Line S has 7 underground stations and a maximum overburden of 21 m.

Reference should be made to the incorporation of an existing old railway tunnel in the Porto Metro network, the Lapa tunnel, as well as the construction of the auxiliary line J excavated by the NATM.

Figure 1 shows the layout of Porto Metro. Lapa tunnel is located near Trindade station.

3.2. Geotechnical conditions

The underground part of Porto Metro was excavated in medium grained two mica granite, which is characterized by deep weathering showing a marked heterogeneity. The underground works cross six grades or weathering alteration ranging from sound granite W1 to residual soil W6, established in accordance with ISRM and Geological Society of London (Miranda et al., 2005; Babendererde et al., 2006).

The rock and soil masses were typified in several geomechanical groups. G1, G2, G3, G4 and G5 were used for rock masses, while G6 and G7 were used for soil formations, respectively residual soils and sediments.

The parameters of the geomechanical groups that were determined through a statistical analysis are presented in Table 1 (Russo et al., 2001). The parameters

Figure 1. Porto Metro network. Location of Lapa tunnel.

Table 1. Geomechanical groups properties.

Group	W	F	Discontinuity condition	GSI
G1	W1	F1-F2	d1-d2	65–85
G2	W2	F2-F3	d2-d3	45–65
G3	W3	F3-F4	d3-d4	30–45
G4	W4	F4-F5	d4-d5	15–30
G5	W5	F5	d5	<20
G6	W6	na	na	na
G7	na	na	na	na

W- weathering degree; F – fracturing degree in accordance with ISRM classification; classes of discontinuity conditions corresponds to the GSI surface conditions.

can also be calculated by a KBS system named GEOPAT (GEOmechanical Parameters for Tunnelling) developed in order to obtain the main geomechanical parameters in both rock and soil masses and also heterogeneous formations in a more rigorous way (Miranda et al., 2005).

3.3 Tunnel J

An auxiliary underground line, designated as tunnel J, was constructed near the Lapa tunnel as indicated in Figure 2 with an extension of 335 m. The average depth

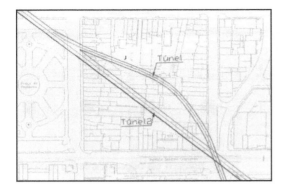

Figure 2.   Location of tunnel J.

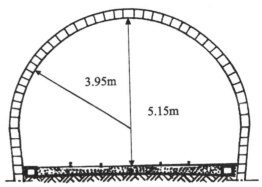

Figure 4.   Masonry cross section of Lapa tunnel.

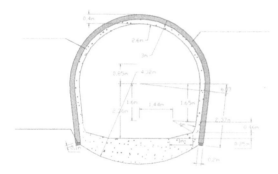

Figure 3.   Section A for tunnel J.

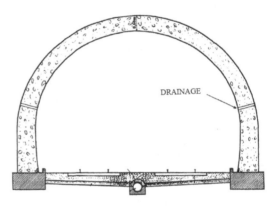

Figure 5.   Croncrete cross section of Lapa tunnel.

is about 21 m, crossing the existent railway tunnel at km 0+066.4 at a distance of 6 m between both tunnels.

For the excavation the NATM method was applied using a sequential and an observational method. Different types of supports were used, mainly shotcrete and bolts (Pimenta et al., 2003). In general due to the small section involved only one step excavation was used.

In general only small buildings, with one to three levels existed at the surface. However, there were meduium sized buildings located in certain areas.

In the design three independent sections were considered (A, B and C), with different geometric and support characteristics. Figure 3 illustrates the geometry and supports for section A.

Each section was associated to different geomechanical conditions, characterized by type 1 and type 2. The section for type 1 was associated to a G2 and(or) G3 rock mass above the tunnel, consisting of primary support of 15 cm thick shotcrete reinforced with fibre or wire mesh and a secondary support of 20 cm thick concrete. This support type was adopted on section C. Section type 2 was considered when G4 or G5 rock occurred above the tunnel. The primary

support consisted of a metallic profile IPE-160 and shotcrete with 20 cm thick reinforced with wire mesh or fibres (section A and B).

4   LAPA TUNNEL

4.1   *Historical data and main characteristics*

The construction of Lapa tunnel was initiated in 1930.10.27 from the Trindade side and in the following day from the side of Boavista station. The tunnel started to be explored in 1938 when Trindade station was open to the public. Figure 4 presents a cross section of the tunnel with a masonry lining and Figure 5 a cross section with a lining in concrete near Trindade station.

The tunnel has a straight alignment with a small inclination. The length is about 494 m and the level difference between the beginning and the ending of the tunnel is 8 m. The tunnel has two rail lines and the cross section has a circular arch with a radius of 3.95 m. The walls 2.24 m height are slightly inclined.

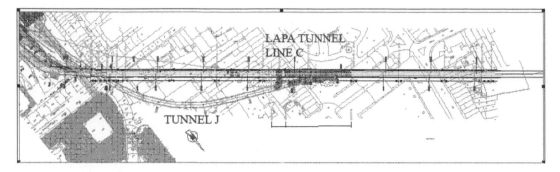

Figure 6. Plan of Lapa tunnel and location of J line.

The maximum overburden is 25 m and the minimum 5 m. At the surface buildings and other infrastructures exist.

LNEC conducted inspections before the rehabilitation works (Sousa et al., 2003). It was possible to conclude that almost all the tunnel presented the occurrence of humidity and important defects in the concrete on the Trindade side. Also several situations of opened joints were detected. In accordance to a methodology used by this institute it was considered the tunnel needed to be rehabilitated. The works should be programmed in function of the future use of the tunnel.

4.2 Geotechnical characteristics

Figure 6 presents a plan of the tunnel as well as the auxiliary line J previously referred.

The Lapa tunnel was excavated in a two mica grained granite. The formations varies from a W4 weathered granite to a highly and decompose granite W4-5.

Boreholes were executed in several sections of the tunnel that permitted to verify the state of deterioration of the lining, as well as the surrounding rock mass (Fig. 7). Figure 8 illustrates one the observed cross sections.

4.3. Description of the rehabilitation works

Figure 9 presents a plan with the major rehabilitation works performed at Lapa tunnel.

The rehabilitation of the Lapa tunnel included five types of works.

Type 1 rehabilitation works consisted of the execution of a concrete invert with a thickness of 50 cm between Pk 0 + 114.87 and Pk 0+150.00. These works were necessary in order to guarantee almost zero settlements at Lapa tunnel during the excavation of the tunnels for J and S lines.

Type 2 works were characterized by the application of grouting in the contact rock mass – support.

SECTION A
PK 0 + 137.23

SECTION A
PK 0 + 211.85

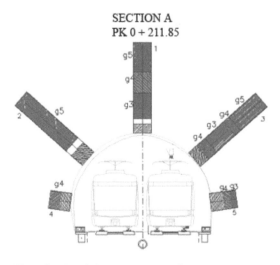
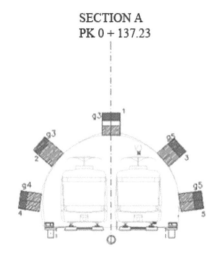

Figure 7. Boreholes at two cross sections.

66

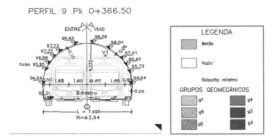

Figure 8. Surveying of the existing tunnel.

Type 3 works was performed in order to rehabilitate the surface of the lining, necessary due to the fact of the deterioration thickness of the lining being larger than 5 mm. In these areas a part of the lining was demolished and afterwards 25 cm thick shotcrete was applied.

Type 4 works consisted of the application of watertight system.

Figure 9 shows the location of Lapa and J tunnel. Figure 10 a typical cross section the different rehabilition type of works.

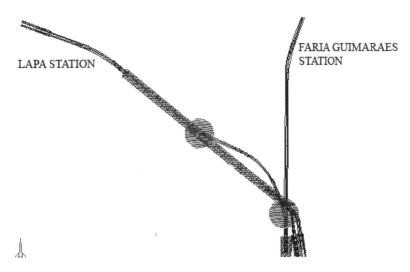

Figure 9. Plan with the location of Lapa and J tunnels.

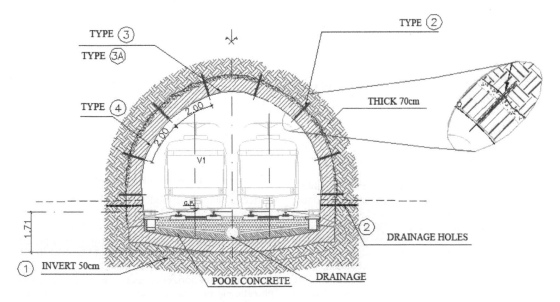

Figure 10. Section type with the different rehabilitation works.

Table 2. Geomechanical characterization of rock masses.

Group	E (GPa)	ν	m_b	s	σ_c (MPa)
G2	8.7	0.15	2.7	4.7e-3	52.5
G3	2.1	0.15	1.35	5.87e-4	22.5
G4	0.7	0.15	0.84	1.41e-4	10
G5	0.17	0.2	0.55	4.03e-4	2

Table 3. Sections considered in the analysis

Section	Relative position
B1	0+285.7 m
B2	0+266.7 m
C1	0+90.4 m
C2	0+66.4 m
C3	0+39.86 m

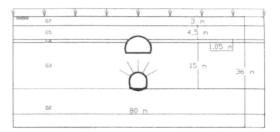

Figure 12. Conceptual model for section C2.

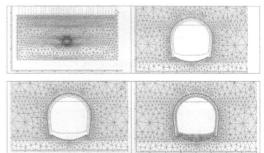

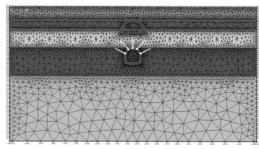

Figure 13. FE model for section B1. Different sequences.

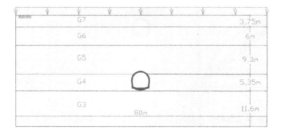

Figure 11. Conceptual model for section B1.

The geomechanical characterization of granitic formations are presented in Table 2.

5 STRUCTURAL ANALYSIS OF THE LAPA TUNNEL REHABILITATION WORKS

5.1. *Structural behaviour of the tunnel*

The structural analysis of the tunnels, in lines C and J, was done using the finite element software *Phase2* from Rocscience.

The three sections analyzed are presented in Table 3.

Section B1 of the tunnel J, represented in Figure 11, is located in geomechanic zones G4 and G5, that corresponds to weak formations. The load action considered to represent the existence of buildings at the surface was adopted to be equal to 25 kN/m². Sections B2, C1 and C3 intersected only tunnel J and the loads at surface were considered to be equal to 40, 50 and 10 kN/m², respectively. The geomechanical characteristics here were better.

Figure 14. FE model for section C2.

Section C2 considered the presence of Lapa tunnel and tunnel J, being the load at surface considered equal to 10 kN/m². As illustrated in Figure 12, Lapa and J tunnels intersected formations G3, G4 and G5.

5.2 *Numerical models adopted and results*

Different numerical models were built for the sections referred at Table 3. FEM models using software Phases2 from Rocscience are illustrated at Figures 13 and 14. For more details on the excavation sequence adopted see Pimenta et al. (2003) and Martins (2006).

The value adopted for k was equal to 0.8 (ratio between in situ horizontal and vertical stresses). The phreatic level was considered not affecting the domain in study.

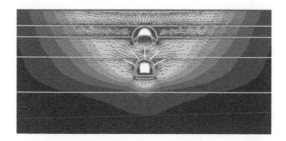

Figure 15.   Total displacements at stage (05).

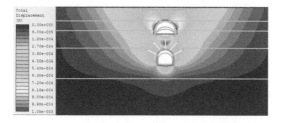

Figure 16.   Total displacements at stage 06 (final stage).

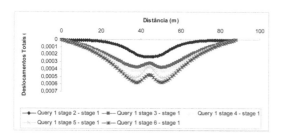

Figure 17.   Settlements at surface (88 m).

Figures 15 and 16 present some of obtained total displacement results for the model of section C2,. When the existent support is removed, major settlements occurred, with a maximum of 1.08 mm. Figure 17 shows to total settlements at the surface and Table 4 the maximum values occurred. Safety factor values were in general lower near the weakest formation and in the vicinity of the excavations.

Empirical methods were also considered in order to assess the risk of building damage due to the tunnelling namely those proposed by Cording and O'Rourke (Pimenta et al., 2003). Table 5 illustrates maximum displacement (δ_{vmax}), distortion angle (β) and horizontal distance of the inflexion point (i_x).

For the relevant section of the two tunnels, section C2, a particular analysis was made. The maximum displacement was within the imposed limits and therefore no special problems at surface occurred.

Table 4.   Maximum settlement values.

Maximum settlement (mm)	6.0
Maximum horizontal displacement (mm)	0.45
Maximum displacement observed	0.72

Table 5.   Application of empirical methods of damage evaluation.

Section	δ_{vmax} (mm)	i_x (m)	β
B1	6.21	11	3.42*10-4
B2	3.01	11	1.16*10-4
C1	0.54	11	2.95*10-4
C3	0.87	12.8	4.11*10-5

6   CONCLUSION

The importance of the rehabilitation works in old railway tunnels was emphasized. A case study was analysed related to the Lapa tunnel in the city of Porto that was recently incorporated in the Porto Metro network. The rehabilitation works for the tunnel are described as well as numerical analyses performed.

REFERENCES

Asakura, T. (2005). ISRM Commission on Maintenance and Repair of Underground Structures in Rock Masses, Brno.
Babendererde, S., Hoek, E., Marinos, P., Cardoso, A. 2006. Geological risk in the use of TBMs in heterogenious rock masses – The case of Metro do Porto. Ed. Campos e Matos, Ribeiro e Sousa, J. Kleberger & P. Pinto, pp.41–51.
Eraud, J. (1982). Rehabilitation of railway tunnels and repair techniques (in French). SNCF, Paris.
Metro do Porto (2002). Tunnel J.- Metro do Porto. Report, Porto.
Miranda, T. (2003). Contribution to obtain geomechanical parameters to modeling underground works in granite formations (in Portuguese). MSc thesis, University of Minho, Guimarães.
Miranda, T.; Correia, A.G.; Sousa, L.R. 2005. Determination of geomechanical parameters using a KBS system and application to an underground station. ISRM Symposium, Moscow, pp 218–225.
Pimenta, M.; Bastos, T.; Vaz, L.G. (2003). Tunnel in rock from Metro do Porto (in Portuguese). University of Porto, Report, Porto, 129p.
Russo, G. et al. (2001). A probabilistic approach for characterizing the complex geologic environment for design of the new Metro do Porto. Proc. Of the AITES-ITA World Tunnel Congress, Milano, pp. 463–470.
Silva, C. (2001). Safety control of railway tunnels. Development of support and KBS methodologies (in Portuguese). MSc thesis, University of Porto, Porto.
Silva, C.; Sousa, L.R.; Portela, E. (2001). Development of methodologies to support the safety control of old railway

tunnels. AITES-ITA 2001 World Tunnel Congress, Milano, pp. 757–766.

Silva, C.; Sousa, L.R.; Portela, E. (2001). Development of methodologies to support the safety control of old railway tunnels. AITES-ITA 2001 World Tunnel Congress, Milano, pp. 757–766.

Silva, C.; Sousa, L.R.; Portela, E. (2002). Behaviour analysis of the São Bento tunnel: application of a knwledge based system. Symposium NARMS-TAC 2002, Toronto, pp. 1507–1513.

Sousa, L.R.; Jorge, C.; Vieira, A. (2003). Inspection and safety control of Portuguese railway tunnels. Trindade, São Bento, Seminário II, China II, Tamel, Caminha, Régua and Bagaúste tunnels (in Portuguese). LNEC Report 396/03, Lisbon, 188p.

Sousa, L.R.; Miranda, T. (2006). Porto Metro Underground Structures. Rock Mass Characterization and Modeling. ISRM Seminar, Lisbon.

Sousa, L.R.; Oliveira, M.; Lamas, L.N.; Silva, C. (2004). Maintenance and repair of underground structures: Case studies in Portugal. Symposium EUROCK 2004, Salzburg, pp. 355–360.

Underground Works under Special Conditions – Romana, Perucho, Olalla (eds)
© 2007 Taylor & Francis Group, London, ISBN 978-0-415-45028-7

Evaluation of the deformation modulus of rock masses using RMR: Comparison with dilatometer tests

José M. Galera,
Geocontrol, S.A. - UPM, Madrid, Spain

M. Álvarez
Geocontrol, S.A., Madrid, Spain

Z.T. Bieniawski
Bieniawski Design Enterprises, Arizona, USA

ABSTRACT: This paper presents the result of comparisons between the modulus of deformation obtained from dilatometer tests and the geomechanical quality of the rock mass using the RMR classification and the basic intact rock properties such as the uniaxial compressive strength and Young's modulus.

The first step was to compare the dilatometer modulus with RQD and RMR. Subsequently, it has been decided to use the RMR without considering the lithology, as the differences where found insignificant.

The second step was to scrutinize the data, excluding those with the following limitations: Weathering grade $\geq$ IV and dilatometer modulus $\leq 0.5\,\text{GPa}$. Also in those cases in which $Em \leq 10\,\text{GPa}$, 15 points were added to the value of RMR because an undrained modulus was being considered.

Excluding any data with anomalous ratios, the final database consists of 436 cases in which known values of Em, RMR, σ_c^i and E' are considered reliable.

With this database several correlations were investigated to estimate rock mass deformability improving on the existing criteria of Bieniawski (1978), Serafin-Pereira (1983), Nicholson-Bieniawski (1990) and Hoek (1995). The results were presented at ISP5 Int. Symp. (Galera et al, 2005). A new relation between RMR and E_m/E^i is recommended:

$$E_m = E_i \cdot e^{(RMR-100)/36}$$

Representing a useful tool for the estimation of rock mass deformation modulus.
Finally in the paper, the new relations are proven using data from two main civil works.

1 INTRODUCTION

The main purpose of this paper is to present a state of the art evaluation of the rock mass deformability and to present the results of comparisons between the modulus of deformation obtained from dilatometer and pressuremeter tests and the geomechanical quality of the rock mass, using the RMR classification. In addition, intact rock properties such as uniaxial compressive strength and Young's modulus are discussed.

2 SOME CONSIDERATIONS OF THE SCALE EFFECTS IN ROCK MASSES

One of the considerations of scale effects in rock masses was by Hoek and Brown (1980) where the strength of the rock mass was estimated by means of the value of the RMR.

Also, the ISRM organized a work group for the investigation of the scale effects in rock masses concerning strength, deformability, joint properties, permeability, and even in situ stresses. The results of these studies were presented in two Workshops, at Loen (1990) and Lisbon (1993).

In particular, concerning the scale effect in rock mass deformability, Pinto de Cunha and Muralha (1990) showed the effect of the volume involved in the test of the deformation modulus measured.

Figure 1 shows this phenomenon, where LAB are laboratory tests, BHD are Borehole Dilatometer Tests and LFJ are Large Flat Jack Tests. Two different ideas can be derived from this figure. One is that the bigger

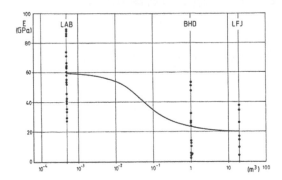

Figure 1. Deformability modulus vs. tested volume (Pinto da Cunha & Muralha, 1990).

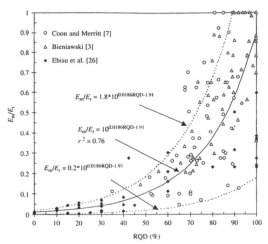

Figure 2. Relationships between RQD and Em/Er (Zhang and Einstein; 2004).

the volume involved in the test, the lower the modulus. The second is that the bigger the volume, the smaller the variability of the results.

3 EMPIRICAL EVALUATION FROM GEOMECHANICAL CLASSIFICATIONS

It is clear that in situ methods are the best approach to predict deformability of rock masses. However, in situ tests are relatively expensive and not always provide reliable results due to several reasons.

Rock mass deformation modulus estimation by correlations with geomechanical classifications appeared as a traditional tool in rock mechanics since Bieniawski (1978) and his RMR index.

Subsequent correlations have included RQD (Gardner, 1987; Kayabasi et al., 2003; and Zhang and Einstein, 2004), Q system (Barton, 1983; Grimstad and Barton, 1993), and RMR (Serafim and Pereira, 1983; Nicholson and Bieniawski, 1990; and more recently, Hoek et al., 1995).

Currently, three different correlations using Q, RQD and RMR are used:

a) Q and rock mass deformation modulus
Barton (1983) and Grimstad and Barton (1993) provided a study with several geophysical borehole measurements obtaining the following relations:
$Q = 10(V_p - 3.5)$ with Vp in km/s, and concluding

$$E(GPa) = 25 \, LogQ \qquad (1)$$

although in other projects $E = 10 \, LogQ$ was found more suitable.

b) RQD and rock mass deformation modulus
Gardner (1987) proposed the following expression,

$$E_m = \alpha_E \cdot E_i \qquad (2)$$

where $\alpha_E = 0.0231 \cdot RQD - 1.32$ (≥ 0.15). This method was used by the AASHTO (American Association of State Highway and Transportation Officials).

More recently Zhang and Einstein (2004) recommended the following relations:

$E_m/E^i = 0.2 \cdot 10^{0.0186RQD-1.91}$ (Lower bound)
$E_m/E^i = 1.8 \cdot 10^{0.0186RQD-1.91}$ (Upper bound)
$E_m/E^i = 10^{0.0186RQD-1.91}$ (Mean)

These expressions are shown in Figure 2. Note the large scatter.

c) RMR and rock mass deformation modulus
The first correlation between RMR and rock mass deformation modulus was proposed by Bieniawski (1978), as

$$E_m(GPa) = 2 \cdot RMR - 100 \quad \text{(For RMR} \geq 50) \qquad (3)$$

Later, Serafim-Pereira (1983) proposed the more known expression,

$$E_m(GPa) = 10^{\frac{(RMR-10)}{40}} \qquad (4)$$

Figure 3 shows graphically both expressions and their comparison.

Nicholson and Bieniawski (1990) derived the following relation considering not only RMR but also the Young's modulus of the intact rock E_i:

$$\frac{E_m}{E_i} = \frac{1}{100} \cdot \left(0.0028 \cdot RMR^2 + 0.9^{\frac{RMR}{22.82}} \right) \qquad (5)$$

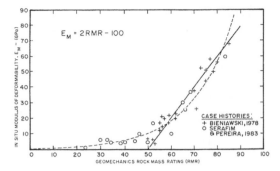

Figure 3. Correlation between the in-situ modulus of deformation and RMR (Bieniawski, 1989).

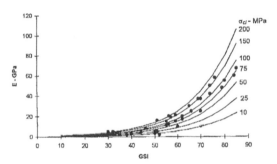

Figure 4. Proposed relationships between GSI or RMR with the intact rock strength (αci) and in situ modulus of deformation Em (Hoek et al., 1995).

More recently Hoek et al. (1995) suggested a correction to the Serafim-Pereira expression, using a factor of $\sqrt{\sigma^i_c(MPa)/100}$, and interchanging GSI (Geological Strength Index) with RMR, as follows

$$E_m(GPa) = \sqrt{\frac{\sigma^i_c(MPa)}{100}} \cdot 10^{\frac{(GSI-10)}{40}} \qquad (6)$$

Figure 4 shows graphically this Hoek et al. (1995) relation.
Finally, Hoek and Diecrich (2006) suggested the following equation:

$$E_m = (MPa) = 100.000 \left(\frac{1 - D/2}{1 + e^{((75+25D-GSI)/11)}} \right) \qquad (7)$$

and also

$$Em(MPa) = Ei \left(0,02 + \frac{1 - D/2}{1 + e((75 + 25D - GSi)/11)} \right) \qquad (8)$$

considering the value of the intact modulus.

The use of RMR and not GSI is strongly recommended because GSI introduces more empirism in a classification that itself is empirical, as was stated in a recent review by Palmström (2003) who warned as follows *"Visual determination of GSI parameters represents the return to quality descriptions instead of advancing quantitative input data as in RMR, Q and RMi systems. GSI was found mainly useful for weaker rock masses with RMR<20.*

As GSI is used for estimating input parameters (strength), is is only an empirical relation and has nothing to do with rock engineering classification".

4 NEW CORRELATIONS BETWEEN RMR AND ROCK MASS DEFORMATION MODULUS

4.1 Database

The information presented here is derived partially from bibliography (Bieniawski, 1978; Serafim-Pereira, 1983; and Labrie et al. (2004)) but mainly from pressuremeter and dilatometers measurements made by Geocontrol during the last decade.
The amount of available data classified by its lithology is the following:

- igneous rocks: 270
- metamorphic rocks: 108
- detritic sedimentary rocks: 175
- carbonate sedimentary rocks: 101
- bibliography: 48

This represents 702 data in which the E_m from in situ tests, RMR and RQD are known.
In 123 of these data also the uniaxial compressive strength (σ^i_c) and Young's modulus of the intact rock (E_i) are also known.
Figures 5a, b, c and d show the available data classified by the lithology. This classification is based on the ISRM and Goodman lithological classifications of rock masses.
Figure 6 shows all the data jointed in the same graph and it can be observed that the differences due to the lithology are insignificant.
The first objective has been to compare the pressuremeter and dilatometer results, which represents the rock mass modulus Em, with RQD and RMR. In Figure 7 a and b both comparison are shown.
It is evident that RMR provides a better trend of the data, since RQD is only one of the five major components of the RMR classification.
This figure clearly shows that RMR is more reliable to estimate the deformation modulus than RQD alone by providing a lesser scatter of data.

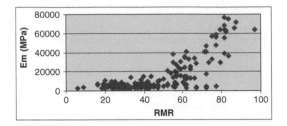

a   Igneous rocks

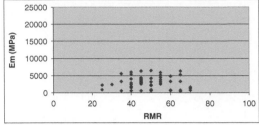

b   Metamorphic rocks

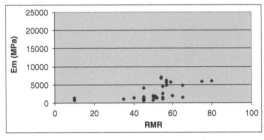

c   Sedimentary carbonate rocks

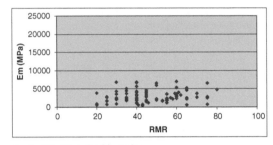

d   Sedimentary detritic rocks

Figure 5.   Database according to the lithology.

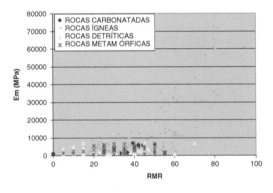

Figure 6.   Relation Em (MPa) vs.RMR according to the lithologies.

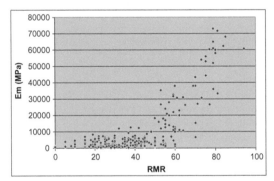

Figure 7a.   Modulus of deformation vs. RMR.

4.2   *Analysis of the data*

The second step has been to scrutinize the data, excluding those with the following limitations:

– Weathering grade bigger or equal than IV.
– Pressuremeter or Dilatometer modulus lesser or equal than 0.5 GPa.

The reason for this filter is to remove data with a "soil behaviour" in which the application of RMR classification is inappropiate as not constituting a "conventional" rock mass.

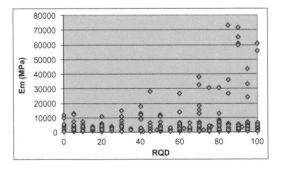

Figure 7b.   Modulus of deformation vs. RQD.

Also in those cases in which $E_m \leqslant 10\,\text{GPa}$, 15 points were added to the value of RMR because a drained modulus was considered.

Celada et al. (1995) analyzed the relation between drained and undrained modulus as:

$$\frac{E_u}{E} = \frac{3(1-v)Kw + E \cdot n}{2(1+v)(1-2v)Kw + E \cdot n} \qquad (9)$$

where Kw is the balk modulus of the water and n is the porosity. From this relation the following can be concluded:

– If E is bigger than $10\,\text{GPa}$, $E_u/E \simeq 1$ and no significant difference exists between both modulus.
– If E is smaller than $10\,\text{GPa}$ and with a drained Poisson's ratio of 0.3, $E_u/E \simeq 1.15$, so the undrained modulus is around 15% higher than drained modulus.

Finally, the third step has been to perform a sensitivity and quality analysis of data, using the following criteria:

– comparison E_i vs. σ_c^i.
– comparison E_i vs. E_m, and
– comparison E_m/E_i vs. RMR.

Excluding any data with anomalous ratios, the final database consists of:

– 427 cases in which E_m and RMR are considered reliable.
– 98 cases in which E_m, E_i, σ_c^i and RMR are considered reliable.

4.3 Discussion

With these data, several correlations have been investigated to estimate rock mass deformability by improving on the existing relationships described in section 3.

Experience shows that with the current correlation usually the deformation modulus Em estimated is higher than the modulus measured by means of borehole expansion tests such as pressuremeters and dilatometers.

Two new different types of relations are proposed:

– without considering Ei values
– including Ei values

In the first case also the values of σ_c^i are included using this expression:

$$\sigma_m = \sigma_c^i \cdot e^{(RMR-100)/24} \qquad \text{(Kalamaras and Bieniawski, 1995)} \qquad (10)$$

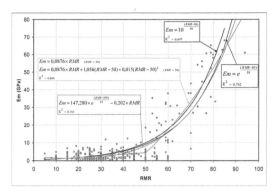

Figure 8. New correlations between RMR and rock mass deformation modulus E_m.

In all the cases the coefficient of regression r^2 has been calculated as follows:

$$r^2 = 1 - \frac{\sum(y_i - y_i')}{\sum(y_i - \overline{y})^2} \qquad (11)$$

where y_i is the value of E_m, $\overline{y}$ is the mean and y_i' are the adjusted values.

a) New correlations between RMR and rock mass deformation modulus

Figure 8 shows all the new correlations considered and also the Serafim-Pereira expression.
i. considering $\sigma_m^i = \sigma_c^i \cdot e^{(RMR-100)/24}$ it is derived:

$$E_m(\text{GPa}) = 147.28\frac{\sigma_m}{\sigma_c^i} - 0.202 \cdot RMR \qquad (12)$$

The coefficient of regression, r^2, obtained is 0.765, that is higher than the one obtained in the regression of the data following Serafim-Pereira, namely, a $r^2 = 0.697$.
ii. The second relation is an improvement of Serafim-Pereira, as follows:

$$E_m = e^{(RMR-10)/18} \qquad (13)$$

The coefficient of regression $r^2 = 0.742$, that improves by 10% the estimation of E_m.
iii Finally, following the original estimation, a threshold of RMR = 50 is derived:

$$E_m(\text{GPa}) = 0.0876 \cdot RMR \quad \text{for } RMR \leq 50 \qquad (14)$$

$$E_m(\text{GPa}) = 0.0876 \cdot RMR + 1.056(RMR-50) + 0.015(RMR-50)^2$$
$$\text{for } RMR > 50 \qquad (15)$$

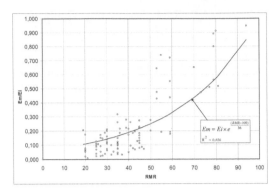

Figure 9. Correlation between RMR and rock mass deformation modulus ratio including E_i.

This above correlation gives a coefficient of regression $r^2 = 0.8$, that improves by more than 15% the estimation of Serafim-Pereira.

b) New correlation between RMR and rock mass deformation modulus including Ei

Figure 9 shows the relation $E_m = E^i \cdot e^{(RMR-100)/36}$.

The coefficient of regression, r^2, is 0.656 which is smaller than that given by the previous correlations but it makes a more reliable estimation as Ei is considered and improves by almost 40% the estimation due to Nicholson and Bieniawski (1990) which gives $r^2 = 0.472$.

5 SUMMARY AND CONCLUSIONS

(1) The Borehole Expansion Tests, mostly Flexible dilatometers, were found to be the best in situ test for the determination of the rock mass deformation modulus.
(2) The empirical Em – RMR correlations present a smaller scatter than the previous correlations Em – RQD.
(3) Several empirical correlations have been studied to estimate rock mass deformation modulus Em. Most of them provide an overestimation of the value Em.
(4) Considering 427 data collected from the published literature and our own data, the best coefficient of regression is obtained considering a threshold of RMR = 50. A linear regression is suggested for values smaller or equal to 50, while a polynomial expression is recommended for values of RMR bigger than 50.
(5) A new relation between RMR and Em/Ei is recommended, considering 98 data. This expression is

$$E_m = E_i \cdot e^{(RMR-100)/36} \qquad (16)$$

representing a useful tool for estimation of the rock mass deformation modulus.

Considering that rock mass strength $\sigma_m^i = \sigma_c^i \cdot e^{(RMR-100)/24}$ and equation (14), it results in the following expression:

$$\frac{E_m}{E_i} = \left(\frac{\sigma_m}{\sigma_c^i} \right)^{2/3} \qquad (17)$$

providing another useful relationship for rock mass characterization.

(6) Nevertheless, the presented correlations should be used realizing that some factors are ignored such as directional effect of jointing.

REFERENCES

Barton N. (1983) Application of Q-system, index tests to estimate shear strength and deformability of rock masses. Proceedings of the International Symposium on Engineering Geology and Underground Construction, vol. I (II), Lisbon, Portugal, 1983, 51–70.
Bieniawski Z.T. (1978) Determining rock mass deformability: experience from cases histories. Int J. Rock Mech and Min. Sc., vol. 15, 237–247.
Bieniawski Z.T. (1989) Engineering rock mass classification: a complete manual for engineers and geologists in mining, civil and petroleum engineering. John Wiley & Sons, 251 pages.
Celada B., Galera J.M., Varona P. (1995) Development of a new calibration and interpretation procedure of pressuremeter test to obtain elastic parameter. The pressuremeter and its new avenues. Ed Balkema, 265–272.
Celada B. Galera J.M., Varona P, Rodríguez A. (1995) Determinación del módulo de elasticidad de formaciones arcillosas profundas. Enresa. Publicación Técnica núm. 01/95.
Celada B., Galera J.M., Rodríguez A. (1996) Ensayos de deformabilidad para la caracterización del terreno in situ. Ingeopres, n. 36, 22–27.
Gardner W.S. (1987) Design of drilled piers in the Atlantic Piedmont. Smith RE, editor. Foundations and excavations in decomposed rock of the piedmont province, GSP ASCE, no. 9, 1987, 62–86.
Grimstad E., Barton N. (1993) Updating the Q-system for NMT. Proc. Int. Symp. Sprayed Concrete, Fagernes, 21 pages.
Kalamaras G.S. and Bieniawski Z.T. (1995) A rock mass strength concept. ISRM International Congress of Rock Mechanics. Tokyo, Japan.
Kayabasi A, Gokceouglu C, Ercanoglu M (2003) Estimating the deformation modulus of rock masses: a comparative study. Int. J. Rock Mech. Min. Sci., 2003, 40:55–63.
Hoek E, Kaiser P.K., Bawden W.F. (1995) Support of underground Excavations in Hard Rock. Ed. Balkema, 215 pages.
Hoek E, Diederichs, M.S. (2006) Empirical estimation of rock mass modulus. Int. Journal of Rock Mechanics & Mining Sciences, vol.43, 203–215.

Labrie D. et al. (2004) Measurement of in situ deformability in hard rock. Proceedings ISC-2 on Geotechnical and Geophysical Site Characterization, Porto, Portugal. Ed. Milpress, 963–970.

Nicholson G. and Bieniawski Z.T. (1990) A non-linear deformation modulus based on rock mass classification. Int. J. Mining & Geol. Eng., vol. 8, 181–202.

Palmström A., Singh R. (2001) The deformation modulus of rock masses – Comparison of in situ tests and indirect estimates. Tunnelling and Underground Space Tech. Vol. 16, 115–131.

Palmström A., Stille H. (2003) Classification as a tool in Rock Engineering. Tunnelling and Underground Space Tech., vol. 18, no. 4, 331–345.

Serafim J.L., Pereira J.P. (1983) Considerations on the geomechanical classification of Bieniawski. Proc. Int. Symp on Eng. Geol. and Underground Construction, vol. I (II), Lisbon, Portugal, 1983, 33–44.

Zhang L., Einstein H.H. (2004) Using RQD to estimate the deformation modulus of rock masses. Int. Journal of Rock Mechanics & Mining Sciences, vol. 41, 337–341.

Underground Works under Special Conditions – Romana, Perucho, Olalla (eds)
© 2007 Taylor & Francis Group, London, ISBN 978-0-415-45028-7

Geotechnical behaviour of shaly rocks crossed by the Pajares tunnels (Section 1)

J.M. Rodríguez Ortiz
Polytechnic University of Madrid

J.M. Gutiérrez Manjón
FCC Construcción, S.A.

J. Ramos Gómez
Iberinsa-Acciona

J.A. Sáenz de Santamaría
Gehma

ABSTRACT: 25 km in length, the Pajares tunnels are located on the Madrid-Asturias high-speed train line currently being built. During the design stage, the geotechnical parameters and behavioural model of the rocky massif most suitable for study were discussed in depth with special attention to the shale formations and, more specifically, to the Formigoso shale formation. High creep pressures in the mid and long-term could be expected although the creep parameters must to be assesses from direct measurements and back analyses. Monitoring results from the already bored tunnels, where depths of up to 550 m been reached, are presented and analysed in the paper.

1 INTRODUCTION

The Parajes Tunnels are formed by two one-way twin tubes, each with a length of 24.6 km and maximum overburden in excess of 1,000 m. Their layout and cross section will allows for trains to circulate at maximum speeds of 350 km/h.

The basic aim of this work was to improve the railway connection between Asturias and the Central Plain of Spain through the Cantábrica mountain range (Figure 1).

2 THE PAJARES TUNNELS, SECTION I

From a constructive point of view, the layout of the Pajares Tunnels has been divided into four Sections.

The design and project corresponding to Section 1 was awarded to the Túneles de Pajares 1 TJV constituted by the construction companies ACCIONA and FCC. The project consists of the following main parts (see Figure 2):

1. The first 10.4 kilometres of each of the two tubes forming the running tunnels. The inner diameter

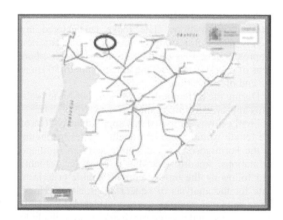

Figure 1.   Location of the Pajares tunnels.

of the tunnel, bored by a TBM is 8.5 m, with an excavated cross section of 77 m².
2. Intermediate access to the tunnels through the Folledo Gallery. This gallery, excavated by the NATM and with a horseshoe section, is 2,091 m in length with an excavated cross section of 75 m²

Figure 2. Layout of the running tunnels and the Folledo access gallery.

Figure 3. General view of the tunnel.

Table 1. Segment concrete strength.

CONCRETE	Allowable working stress (MPa)	TUNNEL LENGTH	
		Metres	(% of the tunnel)
HA-40	19.68	10,186	49%
HA-50	24.60	2,466	12%
HA-60	29.52	2,184	11%
HA-70	36.90	1,792	9%
HA-80	42.16	1,477	7%
HA-90	47.43	1,953	9%
HaA100	52.70	370	2%
HA-105	55.33	282	1%

and a descending gradient of 15–16% down to the tunnels at 8 km from the entrance.

3. There are 25 transversal galleries connecting the two tubes, approximately every 400 m. A typical circular section has been defined for these galleries with an inner radius of 2.50 m (excavated cross section of 32 m²).

The Design for Construction (Section I) was prepared in 2003 by the Túneles de Pajares 1 TJV in collaboration with IBERINSA. Work started that same year and the foreseen completion date was for within 5 years, using the methods indicated in the Basic Project that included conventional methods for the more problematic sections. Approximately 90% of the work has been completed to date (March 2007).

2.1 Geotechnical and construction problems

Section 1 of the Pajares Tunnels is very heterogeneous with a number of different geological formations and frequent alternations. The rocks encountered are mainly shales, limestones, sandstones and quartzites. Approximately 40% of the section includes overburdens of over 450 m, reaching a maximum of 900 m at the end of the section.

During the different design stages, two basic problems appeared as important.

Firstly, it was estimated that approximately 20% of the length would cross bad quality shales belonging to the Formigoso, La Vid, Oville and San Emiliano formations. According to the information available and following the most common empiric criteria to date for the analysis of squeezing problems (Hoek, Barla and Goel criteria), it was considered that all this rock masses, under large overburdens, were likely to develop significant stress-deformation problems in the mid and long term.

Moreover, hydro-geological analyses concluded that water could lead to significant constructive difficulties, as very high discharge flows were foreseeable in some formations (up to 500 litres/sec as a peak discharge). The fact that the tunnels were to be excavated downwards increase the problem as the excavation front would become a blind end.

2.2 Construction method

The construction method chosen for almost the entire Section 1 of the Pajares Tunnels was that of full-section bored by TBM and segmental lining. Two rock shielded machines are being used, which were designed with the possibility of create large gaps (≈15 cm) to allow for high ground deformation before coming into contact with the shield.

The geometric design of the lining corresponds to that of universal rings with a segment length of 1.5 m and a thickness of 0.5 m (Figure 3).

The segments have been defined with different qualities of concrete, depending on the foreseeable pressure of the ground. Table 1 indicates the different types of segments designed for Section 1.

High-strength concrete segments (HA >50 MPa) are foreseen for 39% of Section 1, reaching maximum values of 105 MPa for theoretically more critical areas from a long-term stress-deformation point of view.

3 GEOLOGY

Section 1 runs through the Somiedo-Correcilla Unit of the Pliegues and Mantos Region that, in turn, forms part of the Cantábrica Area, the outermost part of the northern branch of the Iberian Massif (see geological longitudinal profile in Figure 4).

This area is characterised by the lack of metamorphism and the almost inexistence of tectonic foliations. This main structural feature of the region is the presence of a group of large reverse faults with a very low angle in relation to the original stratification and with large northern movements in this area of the section. This type of faults, also known as thrusts, are the main tectonic structures formed during compression caused by the Hercynian orogeny.

Following the settling of the layers, Hercynian compression gave rise to two systems of folds that are almost orthogonal to each other and that affect previous structures. The end result of these stages of folding is that, at present, the initially horizontal thrust is seen vertically or reversed, with the appearance of direct faults. Likewise, these stages of folding lead to the generally vertical position of the strata.

Following the development of the Hercynian orogeny, during the Stephanian period (Late Carboniferous) and also throughout the Permian, Mesozoic and Tertiary periods in the Cantábrica Mountain Range, a series of stages or episodes of faulting were generated, generally known as Tardihercynian or Alpine.

This tectonic is not well-known as it is difficult to map in the field and to learn of its movements and characteristics. These are linear fractures with high dips, significant vertical movements and few horizontal movements and tears that share the rocky massif in large blocks.

In terms of the rock outcrops in the Section, there is a series formed by terrigenous materials (shale, sandstone and quartzite) alternating with calcareous materials (mass limestone, alternance of limestone and marl, marly shale, etc.) that cover a large part of the Paleozoic Era, more specifically from the Cambrian to the Late Carboniferous period (Westfalian). In some areas of the Section, carboniferous materials from the Stephanian period of the Ciñera-Matallana Basin appear on top of these materials in a discordant manner, favouring the Tardihercynian fractures indicated above.

4 GEOTECHNICAL PROPERTIES

4.1 Geotechnics in the design for construction

The numeric calculations in the Design for Construction of Section 1, using the finite difference code FLAC3D focused on the shale formations likely to suffer from squeezing. Due to the tight schedule, no geotechnical investigations were possible when the Design was prepared. So the information from a Basic Design was used along with a detailed revision thereof carried

out by the JV. Table 2 shows the parameters used in the modelling of the more unfavourable sections:

In the calculations, an elasto-plastic behaviour of Mohr-Coulomb type was considered together with a power-law to simulate the delayed or creep strains, as follows.

$$\dot{\varepsilon}_{cr} = A\,\sigma^{-n}$$

where $\dot{\varepsilon}_{cr}$ is the creep or shear strain velocity (only for values above 50% of the strength and null otherwise, according to Rousset, 1988).

"A" and "n", area creep parameters depending on the type of material (in our case A = 1E-27 and n = 2.7)

$$\bar{\sigma} = \text{the von Mises invariant} = \sqrt{3\,J_2}$$

The main difficulty lays in estimating the "A" value. In this case, it was established from some direct measurements in a nearby coal mine, traversing similar shales.

In other sections of Pajares, the creep effects were studied by means of Mohr-Coulomb models, reducing the c and ϕ values in the long term.

The calculations made under creep conditions showed that the equivalent pressure on the segment overburden stood at a maximum of around 30% of the geostatic (a maximum of 29.1% was obtained from the calculations).

In calculations without problems of creep, the pressure of the overburden stood at a maximum of around 20% of the geostatic (a maximum of 17.23% was obtained from the calculations).

4.2 Additional field campaign and new geotechnical characterisation

Following acceptance of the Design for Construction, the JV decided to carry out a new site investigation to complete the geological and geotechnical characterisation along the layout existing at that time. This campaign took place between September 2003 and March 2005.

Table 3 summarises the mean parameters eventually proposed for the different lithotypes expected in Section 1, as a result of the new data.

Two general conclusions were obtained from this new geotechnical and geomechanical characterisation:

1. Significant homogeneity of the results within a given geotechnical group or lythotype.
2. A clear improvement in the strength parameters of most of the lithotypes in relation to those formerly considered. This improvement was evident in the case of the shale, where resistances of around 50% to 100% higher were obtained in Formigoso, Huergas and La Vid shales and 20–25% higher in San Emiliano and Oville formations.

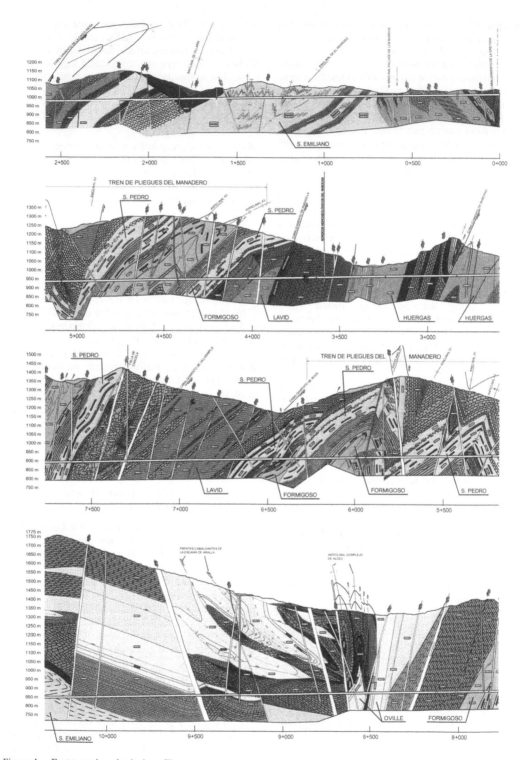

Figure 4.   East tunnel geological profile.

The additional field campaign showed that most of the shaly lithotypes included frequent sandstone and/or quartzite levels, although the shale was always predominant. Logically, the geotechnical properties of these materials depend on the actual percentage of sandstone and quartzite.

5 MONITORING

5.1 *Main tunnels monitoring*

5.1.1 *Pressure cells, extensometers and convergence measurement*

22 instrumented rings have been installed along the completed length (12 in the Eastern Tunnel and 10 in the Western Tunnel). All of the instrumented ring segments include two vibrating wire extensometers that are facing each other and welded to the reinforcements of both faces, as well as a total pressure cell on the outside face.

A total of 138 convergence control sections were also implemented on site, on which the initial readings were made at the time the ring was erected. The introduction of the measuring tapes was adapted to the possibilities of the TBM Back-up.

Figure 5 shows a layout of the ring with the convergence measuring lines and the monitoring devices, corresponding to the Eastern tunnel.

Figure 6 shows the maximum radial pressure values for the different lithologies, measured on each of the instrumented rings in relation to the tunnel overburden. Certain linearity can be observed between the maximum radial pressures and the overburden in the case of shale and sandstone formations.

Most measured values of maximum radial pressures in the shale range between 5% and 6% of the geostatic pressure. Only one of the results, measured in the San Pedro shale formation, exceeds this range reaching 0.75 MPa that means around 10% of the geostatic pressure.

Table 2. Parameters used in the design calculations.

Formation	H (m)	Elastic properties E (MPa)	Elastic properties ν	Plastic properties c' (MPa)	Plastic properties ϕ (°)	ρ (kN/m³)
Formigoso	550	2500	0.3	0.5	23	26
Oville	900	2600	0.3	1.74	28.11	26

Table 3. Parameters assigned following the additional geotechnical campaign.

GEOTECH. GROUP	FORMATION (GEOTECHNICAL LITHOTYPE)	γ (g/cm³)	σ_c (MPa)	σ_T (MPa)	υ	m_i	E_{HOEK} (GPa)[1] Max.	Min.	E_{DIL} (GPa)[2] Max.	Min.
SHALE	Formigoso (FO-5-PIZ)	2.69	24.4	3.2	0.24	6–8	3.704	1.044	1.805	0.890
	S. Emiliano (SE-20-PIZ)	2.62	20.0	3.0	0.21	8	3.353	0.795	1.675	0.645
	Huergas (HU-11-PIZ)	2.66	20.4	3.5	0.36	8	2.137	0.803	0.950	
	S. Pedro (SP-6-PIZ)	2.66	23.3	4.5	0.22	8	4.827	1.144		
	La Vid (LV-8-PIZ)	2.68	22.6	2.3	0.18	6	5.650	1.127	5.000	2.490
	Oville (OV-3-PIZ)	2.67	28.9	3.9	0.19	8	4.031	1.604	1.510	0.900
SANDSTONE AND QUARTZITE	Oville (OV-3-AR)	2.66	96.0	10.9	0.15	17	9.250	3.476	2.620	1.930
	S. Pedro (SP-6-AR)	2.73	93.4	11.4	0.16	17	17.185	3.632	3.160	2.200
	Ermita (ER-15-AR)	2.66	90	9.88	0.15	17	8.956	2.383		
	Barrios (BA-4-CU)	2.60	84	7.34	0.14	20	18.286	2.302		
CONGL. LIMESTONE AND DOLOMITES	Pastora (PA-36-CO)	2.65	42.7	6.1	0.25	18	27.555	2.601	2.375	
	S. Lucía (SL-10-CA)	2.70	46.1	7.9	0.38	9	12.073	5.091		
	Láncara (LA-2-GR)	2.73	62.6	8.7	0.23	9	12.539	5.601		
	Láncara (LA-2-CA)	2.77	75.41	8.6	0.25	10.8	10.320	3.263	4.175	2,450
	La Vid (LV-7-CA)	2.75	75.6	9.3	0.25	12	27.495	3.461		
	Portilla (PO-12-CA)	2.70	68.4	10.4	0.23	9	26.153	4.145	2.200	
	Barcaliente (BC-18-CA)	2.70	66.7	8.3	0.26	8.25	19.367	5.458	3.225	
	S. Emiliano (SE-20-VN-CA)	2.70	66.7	8.3	0.26	8.25	19.367	5.458	3.225	

(1) Modules obtained using the Hoek and Brown formula according to the estimated GSI value.
(2) Modules obtained using dilatometric tests.

Figure 7 shows the evolution of the maximum radial pressures corresponding to the rings installed within the shale lithologies over time. The growth in radial pressure over time is approximate to a logarithm-type law and generally stabilises within the 3 months following ring installation, except in the case of ring A 3819 W where the radial pressure continues to grow after 6 months and in the case of ring A 5225 E that had only been installed for little over a month when the last reading was taken.

This confirms clear creep effects, with long term pressures above those measured at the passing of the machine. Thus, trapping of the machine is possible if any accidental stoppage occurs under great cover.

The radial pressure measured by the different cells installed on any given ring indicates a non-homogenous thrust distribution on the perimeter. Mean cell pressure values on the ring as a whole are around 60% of the maximum value.

Based on the readings of the extensometers, mean compression stress on the most loaded ring of around 7.5 MPa has been deduced. This corresponds to an average axial load of 3,750 kN. These values correspond to ring A 4219 E, which also has a greater radial pressure (0.75 MPa).

On all the rings, the pairs of extensometers indicating the sections with greatest load are measuring a compression of around 14 MPa (rings A 3638 E and A 4219 E).

In general, the mean compression stress on the ring deduced from the extensometers fitted around its perimeter is around 60% that corresponding to the most loaded pair of extensometers.

The mean value of the maximum bending moments deduced from the extensometers on each instrumented ring is around 200 kNm.

The highest value is 580 kNm on one of the rings. On the remaining rings, the maximum values are below 420 kNm.

Maximum shortening values of the horizontal line of around 20 mm are deduced from the convergence

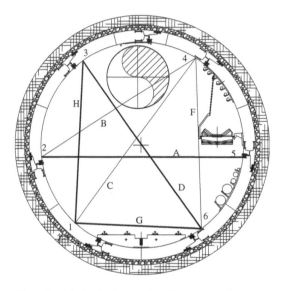

Figure 5. Monitoring layout of the Eastern tunnel.

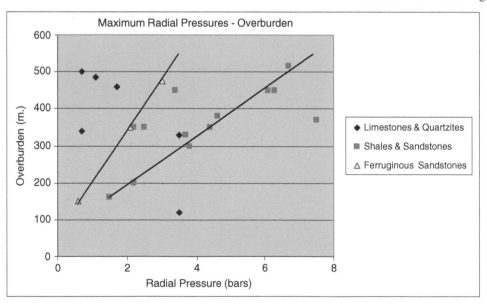

Figure 6.   Maximum radial pressures vs. overburden for different lithologies.

measurements. The Δ_h/ϕ ratio (convergence/diameter) takes values of between 0.1% and 0.2% in 12 sections of control ($\approx$9% total sections), all in shale formations. In the remaining sections, the Δ_h/ϕ ratio is below 0.1%.

5.1.2 *Flat-jack tests*

Two flat-jack tests have been carried out on segment rings to determine their stress at one point of the intrados. These tests were carried out on previously monitored rings and, therefore, it was possible to compare the results obtained with the values from the instrumentation.

5.1.2.1   A3 segment of ring A 4219 E

This test was carried out on the ring with greatest recorded stress along the entire Section 1, according to the results of the monitoring until that time. This ring is below a 350 m overburden of materials from the San Pedro formation, which are formed by thin shale alternances with centimetre layers of quartzite sandstone, although shale predominates.

Table 4 summarises the results of the flat-jack tests, the monitoring measurements and the theoretical estimates made.

There is significant coherence between the results obtained in the flat-jack test and those recorded from the monitoring on segment A3, with relatively uniform compression stress values ($\sigma \approx 7$–8 MPa) and lower, in all cases, than the theoretic value estimated in the Design ($\sigma = 13.56$ MPa). These values mean that the load reaching segment A3 is equivalent to 10% of the geostatic load, i.e. half that theoretically estimated as a design hypothesis.

5.1.2.2   *A1 segment of ring A 5225 E*

This test was carried out on a ring located in an area likely to suffer creep due to the large existing overburden (H $\approx$ 520 m) and the foreseeable bad quality of the shale materials from the Formigoso formation in which it has been placed.

Table 5 shows the results of the flat-jack test, of their comparison with the monitoring measurements on the same ring and the theoretical estimates made.

Due to the fact that certain anomalies were detected on the instrumentation of some segments of the ring (including segment A1), the results of the flat-jack test were analysed by comparing them with the maximum values recorded on the extensometers of the entire ring, which gave compression values of $\sigma \approx 8.5$–10.2 MPa very similar to those of the flat-jack test and, in all cases, below the theoretically estimated value ($\sigma = 35.6$ MPa).

Table 4.   Stress on ring A 4219 E.

Instrumentation		
Flat jack (Segment A3)		$\sigma - 7$ MPa
Extensometer Compression stress	Intrados segment A3	$\sigma = 8$ MPa
	Extrados segment A3	$\sigma = 7.9$ MPa
	Ring average	$\sigma = 7.5$ MPa
Theoretic project hypothesis		$\sigma = 13.56$ MPa

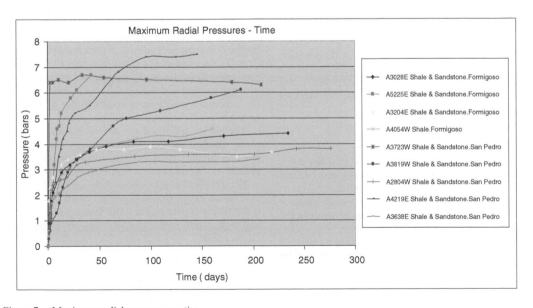

Figure 7.   Maximum radial pressures vs. time.

85

Table 5. Stress on ring A 5225 E.

Instrumentation		
Flat jack (Segment A1)		$\sigma = 9\,MPa$
Extensometer Compression stress	Max. value on Intrados (segment C)	$\sigma = 8.5\,MPa$
	Max. value on Extrados (segment B)	$\sigma = 10.2\,MPa$
Theoretic project hypothesis		$\sigma = 35.6\,MPa$

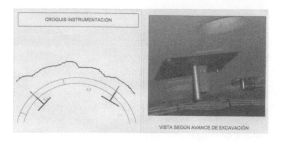

Figure 8. Gap-measurement rods.

Table 6. Final gaps (cm) measured after hardening of the filling.

Rock	Mortar filling				Pea gravel filling			
	C	S	L	I	C	S	L	I
Shale	20,7	15,0	22,3	13,6	18,0	16,2	21,5	17,0
Shale & sandstone	8,0	3,8	22,3	7,8	27,0	12,6	22,2	14,8
Sandstone	28,5	7,7	15,6	12,5	25,7	–	23,3	19,6
Limestone	–	16,9	21,5	11,8	25,7	14,1	23,0	16,2
Quartzite	28,0	20,0	21,0	15,9	23,7	–	20,6	22,0
Heterog.	18,0	–	13,7	12,7	23,6	18,0	22,1	18,3

The maximum values recorded until then meant that the load reaching the ring in a worst-case scenario is somewhat less than 10% of the geostatic load, i.e. one third that theoretically estimated in the Design.

The excavation of the connection of the Folledo Gallery with the Main Tunnels has provided in-depth information on the geo-mechanical quality of the shale and sandstone alternances of the Formigoso formation where ring 5225 is located. Its geo-mechanical quality has been seen to be far superior than that foreseen in the Design for Construction, with RMR values on the face surveys = 50–60. It is therefore logical that the stress-deformation behaviour of these materials be not as unfavourable as initially thought.

5.1.3 Measuring deformations in gaps

The deformation of the ground has been measured on some rings on the perimeter of the excavation using scaled rods fitted in the unfilled gap (Figure 8).

Although the area left unfilled cannot be very large and the ground deformation is limited by the adjoining filled areas, this measurement may help understand the size of ground deformability.

To date, the ring on which greatest deformation was recorded was 5225 in the Western Tunnel, which is at a depth of 520 m in the Formigoso shales and sandstones. Maximum recordings after 12 days of measurements were as follows:

– Segment A3: 0.2 cm
– Segment A2: 3.2 cm

The ground was deformed during the first few days, although this deforming seemed to stop and become stabilised after approximately the tenth day.

More extended measurements have been made of the filled gap, drilling a small hole up to reaching the original ground.

The theoretical gaps of the machines (distance between the bored profile and the outer surface of the segment) under normal conditions are 19–20 cm. This value can be exceeded in case of overexcavation or when the ring displaces itself under selfweight or due to the unbalanced mortar pressure.

After the passing of the shield, the cavity deforms inwards until stopped by the hardened mortar filling or the reaction of the pea gravel. The final thickness of the filling is a measurement of rock deformability as well as the speed of closure of the gap.

Obviously the deformations vary around the ring, although the maximum values do not appear at the same position in each type of rock.

Table 6 shows the final gap measured in both tunnels for the differente rock masses at the crown (C), shoulders (S), middle part of the lateral sides (L) and lower part, close to the invert (I), but not in the invert axis.

It can be observed that, in the case of mortar filling the maximum gap closure, occurs in the shale alternating with sandstones, and mixed ground.

The pea gravell filled gap shows smaller deformations, which could be expected due to the fact that reaction of the gravel is quite instantaneous, while cement mortar takes more than 16 hours to harden.

In no formation the gap has fully vanished , but no conclusion can be drawn in respect of the end closure if the filling is delayed for more than several days. In any case, rapid or instant closure is not to be taken into account.

5.2 Folledo gallery

86 convergence control sections have been established along the approximately 2 km of the gallery. The excavated section is ≈10.5 m wide and ≈8 m high.

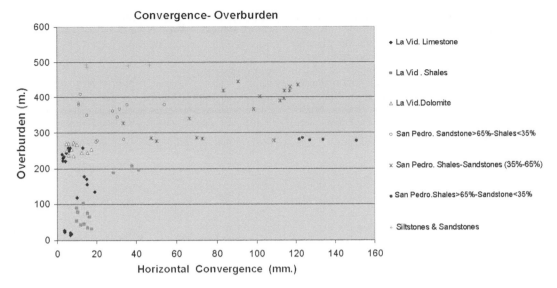

Figure 9.   Horizontal convergence-overburden relationship at the Folledo access gallery.

The gallery was excavated using the drill and blast system.

Four different supporting sections were used. Two of these were used for shale sections with overburdens of over 280 metres and consisted of shotcrete in thicknesses of 25 and 30 cm and a grid of bolts that varied between 1.5 × 1.5 and 1 × 1 meters spacings. One of these sections included TH-29 steel arches, at 1 m spacing.

A plain concrete slab was laid with a thickness of 35 cm, which underwent maximum heaves of 150 mm, thus requiring to increase its thickness and anchoring. In the end section of the gallery, with overburdens of over 400 metres, the flat slab was replaced by an inverted vault with an average thickness of 50 cm.

Figure 9 shows the horizontal convergence values measured along the different lithologies crossed by the gallery.

The maximum value of the $\Delta h/\phi$ ratio is 1.5%, which arises in the sections with predominant (<35% sandstone) San Pedro shale. The non linearity of the relationship can be explained by the use of a temporary support that is too weak for ground pressures.

The sections with sandstone proportions of between 35% and 65% had convergences that were relatively linear to the tunnel overburden. On these sections, the convergence values measures stood at between 0.5% and 1.5% the diameter.

The lowest convergences (<50 mm) were measured on the sections with most sandstone of the San Pedro Formation (>65% sandstone) and on the limestone sections of the La Vid Formation, with relative values of <0.5% the diameter.

6   CONCLUSIONS

Important research work has been devoted to the determination of the long-term pressure and creep effects of the Carboniferous shale formations crossed by the two tubes of the Pajares Tunnel, Section 1.

Theoretical modelling through FLAC[3D] yielded maximum pressures to the order of 30% of geostatic ones. Much lower values (about 10%) could be expected by more competent rock, such as sandstone or limestone.

Current pressure on the lining has been measured by load cells or derived from stress-meters and flat-jack tests. In spite of the low reliability of load cells, it can be concluded that current ground pressure does not exceed 10% the overburden, with many values below 5%.

Creep effects have been also observed, with times to stabilisation of around three months and a significant increase in pressure with respect to those measured at the passing of the machines.

Gap closure measurements behind the shield seem a good tool to detect the risk of development of high pressure on the segmental lining.

Deformation effects were much more significant in the access gallery constructed by the NATM.

ACKNOWLEDGEMENTS

The authors would like to acknowledge and thank everyone to have taken part in developing this Project.

Among these – due to their special devotion – the authors would like to particularly thank Luis de la Rubia and Raúl Miguez from Adif, the Site Manager Julio Santa Cruz, Antonio de Benito from Ineco-Geoconsult for his Technical Assistance, ADIF Consultants Felipe Mendaña and Mario Peláez, Pajares Section 1 TJV Manager Ignacio Muñiz, Site Managers Fernando Fajardo and Jacobo Arnáiz, Tunnel Supervisors Juan Margareto and Pablo Castro, TJV Geologist Marcos Castro and TJV Technical Office Managers Ángel Navarro and Javier Chapado.

REFERENCES

Rousset, G. 1988. Comportement mécanique des argiles profondes-Application au stockage de déchets radioactifs. Thèse. Ecole Nationale des Ponts et Chaussées, Paris.

The selection of a cutter for a tunnel boring machine and the estimation of its useful life

Carlos Lain Huerta
Inforock SL

Pedro Ramirez Oyanguren & Ricardo Lain Huerta
Departamento de Explotación de Recursos Minerales y Obras Subterráneas (ETSI Minas–UPM)

ABSTRACT: In this paper, a procedure to select the best cutter between several ones, i.e., the cutter with the least wear in a determined rock mass, based on the AVS (Abrasion Value Steel) tests, developed by the NTNU, is presented. Once the cutter has been selected, it is also possible to estimate its average duration with the NTNU method. This methodology has been applied to the Guadarrama tunnels and, as demonstrated in this article, interesting results have been obtained. It is worth to highlight the congruency between the actual and estimated life of the cutters during excavation in the granitic rock stretches of the northern tubes of the mentioned tunnels.

1 INTRODUCTION

One of the principal excavation costs in the construction of a tunnel with a tunnel boring machine (TBM) in an abrasive and resistant rock mass is usually the substitution of the cutters.

Nowadays most of the rock masses found in nature can be excavated by means of TBM, but the execution costs can go up by so much so as to result at a disadvantage against the perforation and blasting method. In this article, a practical case for the selection of the cutters for the tunneling machines of the Northern openings of the Guadarrama tunnel, and the prediction of its useful life, will be presented. The materials that were bored in these tunnels were mainly granites and gneisses. The length of the tunnel that was perforated by each machine was about 14,5 km.

The TBM which were used in the Guadarrama tunnels were manufactured by Wirth and Herrenknecht. The first one has a diameter of 9.46 m with 65 cutters and the second one a diameter of 9.51 m with 61 cutters. In Figure 1 a photo of the cutting face of the Herrenneckt TBM can be seen. The cutters can be, according to their position, peripheral, frontal, and central. The first ones, are situated in the periphery of the head and are not perpendicular to the face as are most of the cutters, rather, they form an angle with the tunnel face. The purpose of these cutters is to maintain the correct diameter of the excavation.

Figure 1. Herrenknecht TBM.

The mission of the frontal cutters is to excavate the biggest part of the front between the central and perimetral cutters. The central cutters drill the centre of the section and are, as is the case with most of the previous cutters, perpendicular to the face of the tunnel; that is to the head of the machine.

The peripheral cutters travel a longer distance per revolution of the cutting head and hence are the first ones to wear out. The frontal most external cutters are also usually situated oblique to the face and due to

Table 1. Cutter consumption during the excavation of the Northern opening of the Guadarrama tunnels (Gutierrez Manjon, 2006).

TBM Wirth –lot 3 (northern opening)

	Number	Percentage
Cutters installed at origin	65	1.27
Cutters changed due to wear	4215	82.74
Cutters changed due to blockage	733	14.39
Due to other reasons	81	1.59
Total Numbers of Cutters	5094	100.00
Rings/cutter		1.77
m/cutter		2.84
m^3/cutter		199.57

TBM Herrenknecht –lot 4 (northern opening)

	Number	Percentage
Cutters installed at origin	61	1.09
Cutters changed due to wear	3794	67.78
Cutters changed due to blockage	1393	24.88
Due to other reasons	350	6.25
Total Numbers of Cutters	5598	100.00
Rings/cutter		1.60
m/cutter		2.60
m^3/cutter		181.40

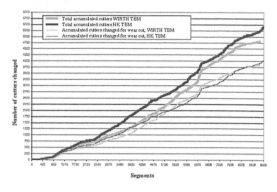

Figure 2. Cutter consumption in the excavations of lot 3 (Wirth TBM) and 4 (Herrenknecht TBM) of the northern openings of the Guadarrama Tunnels (Gutierrez Manjón 2006).

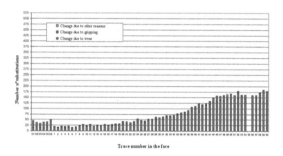

Figure 3. Cutter substitutions frequency. Guadarrama Tunnels (Northern openings) Wirth TBM (Gutierrez Manjón 2006).

their superior distance of rotation, experiment more wear within this group.

Two parts can be distinguished in a cutter: the sleeve, which is usually thermal treated steel, and the cutter in itself. The sleeve is a ring which is put in place while hot into the cutter and which is taken out when it has been excessively worn out due to cutting. This operation can only be undertaken in a machine shop and for this reason the cutters have to be removed completely from the TBM, in order that the sleeve can be changed. One of the most critical elements of the cutter are the bearings which allows it turn around on its axis. When bearings grip, the cutter is not able to rotate and therefore does not wear out uniformly, and for this reason, it is necessary to be replaced. Another problem that can arise is damage to the sleeves due to splinters of steel breaking off the cutter instead of a wearing off, in other words due to fragile breakage.

The bearings of the perimeter cutters suffer blockage more frequently than the others since their position is oblique to the cutter head. They also wear out more since they travel a longer distance over the rock face. These some problems may also arise in the frontal most external cutters which are also oblique to the face and also have to travel a considerable distance.

The central cutters have the longest life-spam, although, due to a small radius of displacement, and

being more confined, receive eccentric forces, and vibrate more.

In Table 1 (Gutierrez Manjon, 2006), the cutter consumption of both TBM during the excavation of the northern openings of the Guadarrama tunnels are detailed. In this table it can be noted that more worn out cutters than gripped cutters have been changed specially with the Wirth TBM. On an average, a cutter had to be changed for each 2.7 m. of advance, that means to say that not even two rings of lining sectors were excavated without having changed any cutter. This high consumption of cutters is due to the granites and gneisses that were bored, that are both resistant and abrasive due to their high quartz content. In Figure 2 (Gutierrez Manjón, 2006) is represented the accumulated cutter consumption in the Guadarrama tunnels. As can be seen, the Herrenknecht TBM consumption has been slightly higher, around 10% more, than that of the Wirth TBM. This figure demonstrates very clearly the importance of the wear in the life of the cutters. Figures 3 and 4 show the number of cutter substitutions of the Wirth and Herrenknecht TBM'S with respect to their position

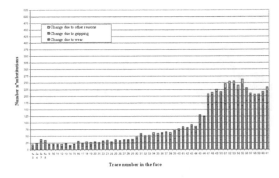

Figure 4. Cutter Substitution frequency. Guadarrama Tunnels (Northern openings) Wirth TBM (Gutierrez Manjón 2006).

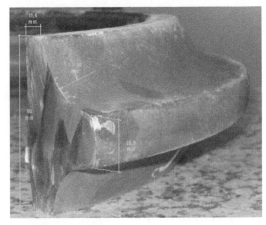

Figure 5. Sector of the sleeve which shows its profile.

(Gutierrez Manjón, 2006). In Figure 4, the drastic increment in consumption in the Herrenknecht TBM can be seen after the cutter that occupies position 47. With the Wirth TBM (Figure 3), the increase in cutter consumption nearing the periferical location is more gradual; and perhaps for this reason, this TBM had less ring consumption than Herrenknecht.

2 DESCRIPTION OF SLEEVES

As previously described, the sleeves or rings of the TBM cutters are manufactured of heat treated steel. In Figure 5, a sector of the sleeve is shown, which allows us the see the profile. This type of profile is known as a constant section since the width of the border cut modifies very little during the wear of the ring. A ring is considered completely worn out when about 40 mm of the radius has been lost. Figure 6 shows a worn out sleeve compared to the circumference of a new one.

The wear of the rings is usually uniform since the distance travelled by every part of them during rotation is approximately the same; although this is not the same when the bearings of the cutters are gripped. In this case, since the ring can't rotate, only a part of it is in contact with the rock, and for this reason, the wear is local and very intense which makes the ring useless very quickly.

The wear of the ring is generally accompanied by a loss of metal splinters and heating of the steel, which could have an influence on its resistance. In Figure7, two sections of the sleeve are shown; one is new, and the other one is worn .As can be seen in this figure, the wear has reached 32,4 mm already, or almost the maximum of 40 mm as previously indicated. The spaces which can be observed in each of the sections of the figure correspond to the testing piece extracted for AVS testing which will be described further on.

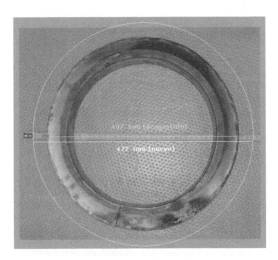

Figure 6. Wear of ring B.

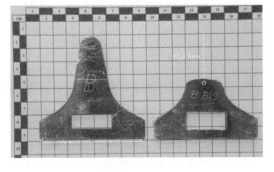

Figure7. Wear of ring B in section.

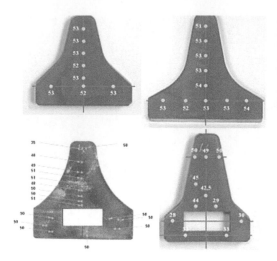

Figures 8 to 11.   Hardness (Rockwell C-150) of rings A, B, D, and H.

Figure 12.   AVS testing machine of the ETSI of Mines of the UPM.

One way to know if the heat treatment of the steel is adequate is to measure the surface hardness of a section thereof. In Figures 8, 9, 10 and 11, sections of the four rings are shown. Further on ,these will be compared in order to find out which one will have a longer cutting life. On these sections, the places where the Rockwell C-150 hardness has been tested are shown. As can be observed in Figure 8, ring A has approximately the same hardness over the entire section (52–53). Figure 9 also shows a fairly uniform hardness, which varies between 51 and 54. On sleeves D and H (see Figures 10 and 11), the hardness is lower and in general less uniform than on the others.

3   AVS AND SJ TESTS

The AVS test (Abrasion Value Steel) lets us determine the resistance to wear of the sleeves. This test measures the abrasion produced by the crushed rock powder of grain size inferior to 1 mm on a piece of steel from the cutter. In order to obtain the testing piece, first of all it is necessary to obtain a 10 mm thick section of the ring and extract a piece as shown in Figures 8 to 11.This has to be obtained without generating any heat(by a cold method), so as not to alter the heat treatment of the steel.

The AVS value is the same as the loss of weight of the testing piece in milligrams after a 10 kg weight is placed on it, and this is worn by the crushed rock powder underneath on a 40 cm plate, within a groove, after 20 revolutions. The form of the test piece is a 30 mm long by 10 mm wide parallelepipe with one of its face curved with a 15 mm radius. The crushed rock powder is fed into the channel groove in the plate by means of a vibrator, and after having passed below the test piece, is removed by a vacuum cleaner so that the rock powder which does the wearing is always renewed.

It is necessary to perform two tests for each rock powder sample in order to obtain the AVS of one rock sample and of one steel cutter. Figure 12 shows a picture of the AVS testing machine belonging to the ETSI de minas of the UPM .This machine has been manufactured with information given by the School of Mines of the University of Trondheim (NTNU) where the test was invented, and has demonstrated its validity in predicting the wear of the TBM cutters.

The SJ or Sievers' Miniature Drill Test, is performed with a small drill, using a 8,5 mm diameter "Widia" bit with a 110° cutting angle, and has the object to estimate the superficial hardness of the rock .The Sievers' SJ value is the drill hole depth on the rock sample, after 200 revolutions of the drill bit, measured in 1/10 mm. The pushing force on the bit is 200 kN.Four drill holes are bored per rock sample in order to obtain the average depth. The machine used by the ETSI of Mines of the UPM to determine the SJ value is shown in Figure13.

Figure 13. SJ. testing machine of the ETSI of Mines of the UPM.

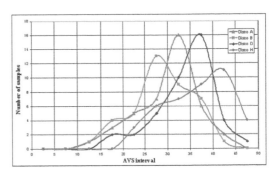

Figure 14. Normal distributions of the AVS value for the four rings.

Table 2. Statistical analysis of the AVS results with cutters A, B, D and H.

Comparasion of cutters					
A y B		A y D		A y H	
$\overline{X}_A = 30{,}23$		$\overline{X}_A = 30{,}23$		$\overline{X}_A = 30{,}23$	
$\overline{X}_B = 29{,}13$		$\overline{X}_D = 34{,}44$		$\overline{X}_H = 38{,}66$	
$\sigma_A = 6{,}47$		$\sigma_A = 6{,}47$		$\sigma_A = 6{,}47$	
$\sigma_B = 6{,}86$		$\sigma_D = 6{,}44$		$\sigma_H = 7{,}08$	
$ERES_{AB} = 1{,}46$		$ERES_{AD} = 1{,}46$		$ERES_{AH} = 1{,}54$	
$\dfrac{DMED_{AB}}{ERES_{AB}} = 0{,}73$		$\dfrac{DMED_{AD}}{ERES_{AD}} = 2{,}88$		$\dfrac{DMED_{AH}}{ERES_{AH}} = 5{,}49$	

Comparasion of cutters					
B y D		B y H		D y H	
$\overline{X}_B = 29{,}13$		$\overline{X}_B = 29{,}13$		$\overline{X}_D = 34{,}44$	
$\overline{X}_D = 34{,}44$		$\overline{X}_H = 38{,}66$		$\overline{X}_H = 38{,}66$	
$\sigma_B = 6{,}86$		$\sigma_B = 6{,}86$		$\sigma_D = 6{,}44$	
$\sigma_D = 6{,}44$		$\sigma_H = 7{,}08$		$\sigma_H = 7{,}08$	
$ERES_{BD} = 1{,}51$		$ERES_{BH} = 1{,}58$		$ERES_{DH} = 1{,}53$	
$\dfrac{DMED_{BD}}{ERES_{BD}} = 3{,}53$		$\dfrac{DMED_{BH}}{ERES_{BH}} = 6{,}04$		$\dfrac{DMED_{DH}}{ERES_{DH}} = 2{,}76$	

4 COMPARISON OF THE RESISTANCE TO WEAR OF VARIOUS TBM RINGS

As previously indicated, the sleeves are the parts that have to be changed most frequently in a TBM, due to wear, and for this reason, in order to compare the duration of the four rings selected in this study, a statistical analysis has been realized ,based on the AVS results obtained on the steel.

The testing has been performed on 40 crushed rock samples obtained from surface core drilling samples. The crushed rock, inferior to 1 mm, in grain size has been obtained from crushing and grinding rocks similar to those that would be bored by the TBM upon the executing of the tunnel. During the advance of the TBM, and during the rock cutting process, rock powder of similar grain size is in contact with the cutter. For this reason, the crushed rock powder is used in the AVS test. However, it is convenient to mention that the TBM cuts the rock in a humid state in which the powder forms a cake due to the water effect, which adheres to the cutter. Perhaps, for this reason, the AVS test should be modified by wetting the crushed rock powder.

As previously mentioned, four rings have been tested (A, B, D, H) The AVS values obtained from these four sleeves have been represented in Figure 14, in which it can be seen that all of the rings have different results: the lowest AVS results came from ring B,then A, after that D, and finally ring H which has worn off most in the tests. The problem planted is to discern if the differences of the AVS results of the rings are due to sampling errors, or methodology, or if the wear resistance of the sleeves is really different. This question is important when it comes to selecting a ring which should have a longer life span. The rings were compared two in two. To do this, the variance of the difference of the arithmetical means of rings A and B was calculated in the first place. This is equal to the typical deviation raised to the square power of the AVS results of ring A divided by the number of tests carried out with this ring subtracting one (39), plus the typical deviation raised to the square power of the AVS results of ring B divided by the number of tests on ring B minus 1 (39).The standard error of the

Table 3. Test performed on granite samples obtained from the northern tubes of the Guadarrama tunnel.

Sample	SJ 1/10 (mm)	S20%	AVS (mg)	DRI	CLI	Qz equiv (%)
W-1	12,1	33,75	41	33,75	8,65	51,5
W-2	19,56	47,79	42,5	50,79	10,27	51,86
W-4	13,86	49,2	40	49,2	9,21	54,23
W-5	14,54	50,19	32	50,19	10,22	46,83
W-5-6	27,43	54,13	28,5	57,13	13,64	54,16
W-6	28,96	54,58	27	57,58	14,22	54,16
W-7	89,91	67,37	32,5	80,37	20,47	46,3
W-8	23,74	61,01	36	64,01	11,79	47,91
W-9	23,19	54,38	33	50,32	12,08	53,44
W-10	15,45	53,34	30,5	53,34	10,65	49,54
W-11	40,93	57,7	40,5	64,38	13,9	50,7
W-14	27,03	49,43	32,5	52,43	12,89	60,11
W-15	25,11	65,29	30,5	68,29	12,84	57,01
W-16	12,9	29,84	28,5	29,84	10,2	49,76
HK-2	9,19	58,05	45,75	58,05	7,46	50,95
HK-3	15,73	50,02	43	53,02	9,4	52,77
HK-4	13,25	42,21	42,5	42,21	8,84	52,12
HK-5	27,38	61,17	33	56,97	12,88	57,38
HK-6	14,03	56,97	33	56,97	9,96	44,64
HK-7	15,31	53,05	29	56,05	10,83	52,34
HK-8	17,58	47,97	28	50,97	11,57	47,3
HK-10	17,99	54,66	29	57,66	11,52	42,77
HK-10b1	28,06	46,83	30	49,83	13,49	58
HK-10b2	9,33	46,83	30	46,83	8,83	45,5
HK-10c	7,6	31,49	34,5	31,49	7,73	39,24
HK-11	21,83	51,66	33,5	54,66	11,74	52,8
HK-12	41,1	43,87	38,5	53,87	14,19	56,7
HK-13	43,59	45,81	30	55,81	15,98	52,84
HK-14	27,14	56,65	35,5	59,65	12,48	51,5
HK-15	35,71	51,36	30,5	61,36	14,71	47,63
HK-17	24,43	47,71	38	50,71	11,68	58,78
HK-17b	12,35	35,07	27	35,07	10,24	54,88
HK-18	36,41	41,78	31	51,78	14,72	56,91
	24,02	**50,04**	**33,84**	**52,87**	**11,80**	**51,59**

differences of the means is obtained as the square root of the variance.

Finally the difference of the average AVS values obtained in the tests is compared with the mentioned standard error.

If this difference is bigger than twice the standard error, it is considered that the two compared rings, that is, A and B, in this case, really have different resistance to wear. If this were not the case, we would consider that there isn't a considerable difference between the two rings, although B has worn less than A. In a similar way, the calculations have been realized with the results of rings A-D, A-H, B-D, B-H. and D-H. The results of all of these calculations are reflected on Table 2. As can be observed on this table, rings A and B don't present significant differences. There are, though, differences between rings A and D, B and D, A and H, and B and H.

In summary, rings A and B have a similar resistance to wear, although ring B wears out less. Ring D and H wear out more than the others. Ring D is of a lesser duration than A and B, and H is the ring that wears out more than any other.

5 ESTIMATING THE USEFUL LIFE OF THE CUTTERS AND COMPARISON OF THE ACTUAL RESULTS

The NTNU method has been used for the estimation of useful life of the cutters installed in the Wirth and Herrenknecht TBM's which have bored the northern tubes of the Guadarrama tunnel. This method is based on the CLI (Cutter Life Index) of the rocks. This is obtained from the AVS and SJ results, which test techniques have been described previously, using the following formula:

$$CLI = 13.84 \left(\frac{SJ}{AVS} \right)^{0.3847}$$

Table 4. Estimated and actual duration of the Wirth cutters (Tunnel 3).

Stretch	Length (m)	Sample	Estimated h/disc	Real
3	482,20	W-1	1,08	1,26
		W-2	1,24	
5	1.159,85	W-4	1,10	2,06
		W-5	1,28	
		W-5-6	1,47	
		W-6	1,35	
6	865,08	W-7	1,87	2,16
7	1.496,27	W-8	1,22	1,93
		W-9	1,23	
		W-10	1,17	
		W-11	1,39	
9	1.361,70	W-14	1,21	1,60
		W-15	1,24	
		W-16	1,26	
			1,29	**1,85**

Table 5. Estimated and real duration of the Herrenknecht cutters (Tunnel 4).

Stretch	Length (m)	Sample	Estimated h/disc	Real
3	479,00	HK-2	0,94	1,01
		HK-3	1,13	
		HK-4	1,1	
5	1.267,18	HK-5	1,37	1,76
		HK-6	1,29	
		HK-7	1,27	
		HK-8	1,4	
6	645,61	HK-10	1,32	1,59
7	1.512,29	HK-10b1	1,27	1,17
		HK-10b2	1,05	
		HK-10c	1	
		HK-11	1,2	
		HK-12	1,33	
		HK-13	1,65	
8	701,68	HK-14	1,42	0,98
		HK-15	1,64	
9	1.441,80	HK-17	1,26	1,26
		HK-17b	1,19	
		HK-18	1,5	
			1,28	**1,30**

The rocks that were bored with the mentioned machines were mainly granites of the Guadarrama Mountains. During the advance of the TBM's, the granite samples were taken from the conveyer belt ,14 of which came from the Wirth (W) machine and 19 samples from the Herrenknecht machine. The tests results of these samples are shown on Table 3. The average values of these tests are as follows:

– Sievers' J value (SJ) = 24.02
– Brittleness (S_{20}) = 50.04
– Abrasion (AVS) = 33.84
– DRI (Drilling Rate Index) = 52.87
– CLI (Cutter Life Index) = 11.80
– Equivalent quartz content of the granite = 51.59

Tunnel 3 was excavated with the Wirth machine and tunnel 4 with the Herrenknecht machine. The tubes that were perforated with the mentioned TBM's have been divided into similar geological stretches. In tunnel 3, stretches 3, 5, 6, 7, and 9, were predominantly granites; whereas, in tunnel 4, the granitical stretches were 3, 5, 6, 7, 8, and 9.

The tubes are separated by only 21 m and for this reason the rocks that were bored by both TBM'S are very similar.

In order to estimate the useful life of the entire set of cutters in the head, measured in hours, 65 in W and 61 in HK, an abacus of the NTNU was used with the CLI value of the granite and a cutter diameter of 432 mm. The life span that was thus obtained has been modified taking into account the following criteria: the diameter and rotation velocity of the head, the number of cutters and the equivalent quartz content of the rock. Dividing this corrected value by the number of disks, the average life of the disk is estimated. Tables 4 and 5 represent the calculated result for each one of the rock samples extracted from the tunnels. From these tables, the estimated average life of the cutters of the Wirth machine has been estimated at 1, 29 hours, and 1,28 hours for the Herrenknecht machine. The coincidence is logical, since the rock that was cut is practically the same in both tunnels.

In these tables, the real average duration of the disks are shown for each one of the stretches into which the tunnel has been divided. This information has been supplied by the constructing firm. As indicated previously, the stretches have similar geological characteristics, and for this reason it is logical to talk about the average life of the disk per stretch. Calculating a weighted average, that is to say, taking into account the length of the stretches, the actual average duration of the disks are obtained when they cut the granite. This data is shown in Tables 4 and 5.

The actual average duration of the Herrenknecht discs were 1.30 hours, and the estimated duration 1.28 hours, which means to say a practically perfect coincidence. The difference in disk life in the case of the Wirth TBM between the actual (1,85 hours) and the estimated (1,29 hours) is really substantial. It is impossible to explain this discrepancy using only the characteristics of the granite, but making an observation of Table 1, it can be seen that with the Wirth TBM, there was less gripping of the cutters and this can justify a lower cutter consumption. Also as mentioned, there is a dramatic increase in cutter consumption nearing the periferical locations in the HK machine.

6 CONCLUSIONS

The investigation described in this article provides us with a method to select the most adequate cutter to excavate a tunnel in a determined rock mass, using a TBM of known characteristics. In this way, the consumption of the cutters will be the least possible, and will decrease the cost and time of execution of the tunnel. With the comparative study of the actual, and the estimated consumption of the cutters, we can deduce that the NTNU method allows us to predict the cutter life of the disk with sufficient exactitude. Of course, there are always factors involved such as: the ability of the machinist, the characteristics of the machine and the cutters, which can change the prediction, which is based only on the rock and steel properties.

REFERENCES

Gutiérrez Manjón, José Manuel (2006). Túnel de Guadarrama, capitulo "Consumo de cortadores de los túneles de Guadarrama". Ed. Entorno Gráfico (p. 283–300).
Bruland, Amund (1998). Hard rock tunnel boring. Advance rate and cutter wear. NTNU.

Field test study on the endurance of early embedded bolts

X.M. Zeng & S.M. Li
Luoyang Hydraulic Engineering Techniques Institute, Luoyang, China

ABSTRACT: In this paper, the coupling effect of stress corrosion and chemical corrosion on the model bolts, which were embedded in Zhucun colliery of Jiaozuo of Henan Province 17 years ago, is researched. The corrosive conditions and the corroded state of the model bolts are introduced. The spot corrosion, pitting corrosion, weight loss, corrosion rate, and strength loss of the model bolts are tested and analyzed. The tests and analyses show that the model bolts' (the exposed steel cylinders') strength loss ratio is about 14%, their diameter loss is about 10%, and their sectional area loss is about 19%. Their weight loss is extremely non-uniform, and the maximum weight loss is about 24.4 times of the minimum. The tensile limit load of the model bolts (the steel bars wrapped in mortar), which had been in service for 17 years, is about 18.4%~22.2% lower than that of the new model bolts.

1 INTRODUCTION

Anchorage type of structure refers to those reinforcement or support structure of rock and soil engineering, such as bolt, cable bolt, soil nail, etc. These techniques were applied early in foreign countries, therefore the endurance problem of these techniques were also studied and understood early in foreign countries [1–5]. Among these reinforcing techniques, bolt has the longest application history. During the end of 1960s to the beginning of 1970s, the technical standards of stratum bolt and the technical regulations of cable bolt had been promulgated in France, Switzerland, Czech and Australia. During the end of 1980s to the beginning of 1990s, the technical manuals of soil nail also had been promulgated in these countries. In these technical standards, the protection of bolt, cable bolt and soil nail in corrosive conditions was well considered, and the design and construction were strictly stipulated. R. Schach et al. in Norwegian Institute of Rock Blasting Techniques edited Rock Bolting – a Practical Handbook in 1975 and the 2nd edition in 1979. A series of books on soil and rock engineering reinforcement were published in Germany in 1980s. For instance, T. H. edited Foundation in Tension Ground Anchors and B. Stillborg edited Professional Users Handbook for Rock Bolting. In these books, the general prevention from corrosion of tensile cable bolt and rock bolt was clearly put forward. ASTM published a set of books, totally 8 books, on describing the corrosive effect on metal material in various up-ground and underground conditions during 1974 to 1981. In these books, the content on natural conditions' corrosion, stress corrosion and anticorrosion measures was very

important. M. J. Turer of England substituted a high-strength, corrosion resisting and economical polyurethane web for steel bar as soil nail in 1999. This web had good support performance. M. J. Turer named this web Permanail (permanent nail), however, he didn't give any data on the service life of this web. R. Eligshause and H. Spieth of Germany had researched the structural performance of plug-in screw-thread steel bar. They pointed out the service life of the steel bar would be reduced from 100 years to 75 years if the drilled hole was not clean and the bonding was bad. However, they didn't give any estimation basis and details. In China, concrete structure endurance has been researched well [6–10], but the endurance and service life of anchor type of structure has been studied little. During July of 1985 to July of 1987, Lei Zhiliang et al. had preliminary studied the service life of bolt. The title of his research subject was Corrosion and Protection Study of Mortar Bolt. He got some good results, but his field sampling was far from universality and indoor test was far from being systematic and deep. Zhang Yong et al. had edited Chinese military standard (GJB3635-99), Technical Specification of Design and Construction of Prestressed Cable Bolt in Rock and Soil Engineering, based on their experience and some foreign standards. In one chapter of the standard, the anticorrosion for the prestressed cable bolt had been prescribed. However, the prescription lacked theoretical and experimental basis. Many protective measures were just based on engineering experience. The design life and residual life were still unable to be predicted. During 1996 to 1997, Zhou Shifeng et al had tested the endurance of cement mortar. The title of his subject was Test Study of Cement Mortar's

Endurance in the Corrosive Condition of Underground Engineering. However, his test was indoor single factor corrosion test [11–12]. Reference [13] points out that the research of security and endurance of anchor type of structure is just in the beginning phase and the endurance problem is very serious in China. Corrosive conditions vary with the different civil engineering. Some research methods and results only can be referenced, but can't be used totally for different engineering.

From the above mentioned, there are a lot of research results of security and endurance of anchor type of structure, but only little in China. However, the report on the field corrosive test of anchor type of structure considering many coupling corrosive factors can been seen neither in Chinese data nor in foreign data.

In May of 1986, a batch of model bolts applied with different grade of prestress was embedded in the Zhucun colliery in Jiaozuo of Henan Province by Luoyang Hydraulic Engineering Techniques Institute. The test periods are 1 year, 5 years and 10 years respectively. After the test of 1 year period of model bolts was finished, the tests of 5 and 10 years period of bolts were not been performed for lack of funds until July of 2003. Under the help of Chinese National Natural Science Fund, all those model bolts were dug out in July of 2003 and well tested. This paper summarizes the test's processes and results.

2 TEST METHOD

As shown in Fig. 1a, the model bolt included a steel bar and a steel cylinder. The material of both the steel bar and the steel cylinder was Q235A steel. The steel bar's diameter was 3 mm. The steel cylinder's outer diameter was 40 mm, and its length was 60 mm. On the middle of the steel cylinder, a small hole was made for grouting. The steel bar went through the cylinder. A nut was tightly screwed on each side of the steel bar with a spanner. In this way, the steel bar was prestressed with tensional stress and the steel cylinder was prestressed with compressive stress. Mortar was injected into the steel cylinder through the small hole. The water-cement ratio and cement-sand ratio of the mortar were 0.55 and 0.40 respectively. The cement grade was No. 325.

The made specimens, totally 59, were carried to the test field and set in the drilled holes, and then the drilled holes were sealed by cement mortar, seeing Fig1.b. After 17 years, this batch of specimens are dug out and tested. The main tested parameters include: (1) the corrosive conditions, (2) the corroded state (3) the strength loss ratio, (4) different factor's influence on the service life.

The steel bar simulated the bolt, which was constructed well and under perfectly chemical corrosion and stress corrosion conditions. The steel cylinder simulated the bolt, which was constructed badly and under serious chemical and stress corrosion conditions. Here, the construction of bad quality refers to the bareness of bolt's partial surface due to insufficient injection of mortar.

The no. of the specimens is given in the Table 1.

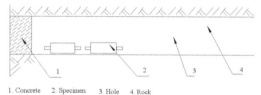

1. Stretching joint 2. End cap 3. Steel cylinder 4. Grouting hole
5. Steel bar 6. Screwed nut 7. Cement paste

a. Specimen

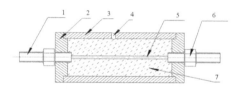

1. Concrete 2. Specimen 3. Hole 4. Rock

b. Setting of the specimen

Figure 1.   Specimen and its setting.

Table 1.   No. of the specimens.

| Mortar | Protective coat | 1 year | | | | 17 years | | | | | | | |
		80 kg		150 kg		80 kg		150 kg		80 kg		150 kg	
Common mortar	None	12	13		43	35	40	44	45	36	37	46	47
Common mortar	Chlorosulfonated polyethylene coatings	5	8	1	2	19	34	3	9	30	39	4	17
Common mortar	Acroleic grout coat	28	29	6	7	14	21	10	11	29	33	20	26
Mortar mixed with additive	None	25	31	15	16	41	50	18	22	32	49	23	24
Acroleic mortar	None	53	60	38	62	52	57	59	61	51	55	54	56

For some historical reasons, the specimens for comparison test were not made then. We have to choose the same type of steel to make the specimens for comparison. These specimens are labeled as 'the new specimens'. The results of comparison test are only used for reference.

3 TEST RESULTS

3.1 *Test of corrosive conditions*

Corrosive conditions of the specimens mainly refer to the temperature, humidity, and contents of various

corrosive matters in the underground water. The test's results are as follows.

(1) Temperature: 19.5°C.
(2) Humidity: 93%~95%.
(3) The contents of various corrosive matters in the underground water are given in Table 2 and Table 3.

3.2 *Test of Corrosive States of the Steel Cylinders*

(1) Pitting corrosion depth of the steel cylinders
The test's results of the pitting corrosion depth of the steel cylinders are given in Table 4.

Table 2. Content of the chemical compositions in the underground water.

Chemical composition	Content
Total acidity	0.318
Free carbon dioxide (CO_2)	140 mg/L
Erosive carbon dioxide (CO_2)	0.00 mg/L
Dissolved oxygen (O_2)	0.30 mg/L
PH value	7.44

Table 3. Content of the ions in the underground water.

	Content of the ions	
Ion	Milligram/liter	Milliequivalent/liter
$Ca2+$	88.18	4.40
$Mg2+$	38.04	3.13
$Cl-$	24.3	0.685
SO_42-	5.28	0.110
CO_32-	0.00	0.00

Table 4. Pitting corrosion depth of the exposed steel cylinders.

Serial number	No. of the specimen	Pitting corrosion depth of the steel cylinder, mm				
		1st	2nd	3rd	4th	5th
1	3	1.26	0.50	0.36	0.44	
2	4	0.63	0.37	0.52		
3	9	0.39	0.12	0.16		
4	10	2.30	2.69	1.38		
5	11	0.95	0.98	1.45	1.00	
6	17	1.73	3.56	2.50		
7	18	0.93	0.62	0.56	0.25	
8	20	0.44	0.35	0.25	0.20	
9	21	3.90	3.10	2.15	3.40	
10	22	0.70	1.50	0.76		
11	23	0.55	0.46	0.38	0.32	0.50
12	24	0.22	0.32	0.36		
13	26	1.12	1.55	2.04		
14	29	1.44	1.37	3.03	2.44	1.35
15	32	3.50	3.15	2.75	1.45	
16	33	1.90	2.00	1.42	1.14	0.70
17	34	1.92	1.06			
18	35	2.78	2.22	1.81	1.60	1.22
19	36	1.07	0.31	0.45	0.57	
20	37	3.45	3.04	3.24	1.85	2.90
21	39	1.36	1.50	2.41	1.74	
22	40	0.90	0.56	1.30		
23	44	1.37	2.70	0.68	0.90	1.34
24	46	2.70	1.31	4.00	2.32	2.2
25	47	1.05	1.63	1.16	1.00	0.62
26	49	0.85	0.45	0.21		
27	50	0.43	0.93	1.13	0.98	
28	51	0.80	1.61	1.15	0.52	0.72
29	52	0.65	0.25	0.15		
30	54	0.25	1.10	1.05	0.75	
31	55	1.03	1.79	0.96	0.91	0.66
32	61	1.62	2.37	1.08	1.43	

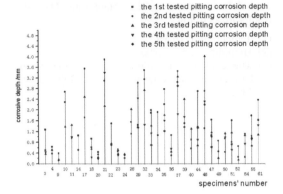

- the 1st tested pitting corrosion depth
- the 2nd tested pitting corrosion depth
- the 3rd tested pitting corrosion depth
- the 4th tested pitting corrosion depth
- the 5th tested pitting corrosion depth

Figure 2. Histogram of the pitting corrosion depths of the exposed steel cylinders.

Fig. 2 is the histogram of the pitting corrosion depth of the steel cylinders.

(2) Weight loss, weight loss ratio and average corrosion rate of the steel cylinders

The tested corroded state of the steel cylinders is given in Table 5. The weight loss, weight loss ratio and average corrosion rate of the steel cylinders are shown in Fig. 3.

(3) Strength loss ratio of the steel cylinders

The results of the strength loss comparison tests between the exposed steel cylinders and the new steel cylinders are shown in Fig. 4.

Table 5. Tested results of the exposed steel cylinders.

Serial number	Specimens' number	Weight loss	Weight loss ratio	Average corrosion rate	Corroded state, area	
		g	%	mm/year	Outside surface	Inside surface
1	3	16.99	0.080	0.072	Badly rust-eaten	80%
2	4	3.36	0.016	0.014	Badly rust-eaten	30%
3	9	17.39	0.082	0.074	Continuously rusty spot	40%
4	10	11.32	0.054	0.048	One pit-corrosion	80%
5	11	18.15	0.086	0.077	Two pit-corrosion	20%
6	14	4.15	0.019	0.017	Two pit-corrosion	80%
7	17	12.63	0.059	0.053	Banded spot-corrosion	70%
8	18	3.32	0.016	0.014	Banded spot-corrosion	60%
9	20	12.73	0.060	0.054	Much rust-eaten	90%
10	21	15.46	0.073	0.065	Bad pit-corrosion	80%
11	22	17.34	0.082	0.073	Uniformly rust-eaten	60%
12	23	8.8	0.042	0.037	Badly rust-eaten	40%
13	24	1.16	0.005	0.004	Slightly rust-eaten	10%
14	26	10.55	0.050	0.045	Two pit-corrosion	10%
15	29	20	0.095	0.085	Bad pit-corrosion	98%
16	32	15.85	0.075	0.067	Badly rust-eaten	
17	33	7.14	0.034	0.030	One pit-corrosion	10%
18	34	3.72	0.018	0.016	Three pit-corrosion	40%
19	35	25.83	0.122	0.110	Much pit-corrosion	80%
20	36	2.54	0.012	0.011	One side rust-eaten	70%
21	37	20.57	0.097	0.087	Bad pit-corrosion	60%
22	39	8.57	0.041	0.036	One side pit-corrosion	10%
23	40	1.66	0.008	0.007	Little spot-corrosion	50%
24	41	8.21	0.039	0.035	Uniformly rust-eaten	5%
25	44	10.4	0.049	0.044	One side pit-corrosion	90%
26	46	19.04	0.090	0.081	Two pit-corrosion	60%
27	47	6.18	0.029	0.026	Banded spot-corrosion	60%
28	49	11.67	0.055	0.049	Little spot-corrosion	80%
29	50	12.69	0.060	0.054	Slightly rust-eaten	60%
30	51	10.85	0.051	0.046	One side pit-corrosion	40%
31	52	7.83	0.037	0.033	Slightly rust-eaten	10%
32	54	7.92	0.038	0.033	One side pit-corrosion	40%
33	55	11.3	0.054	0.041	Much rust-eaten	20%
34	61	9.94	0.047	0.042	Three pit-corrosion	80%

3.3 *Weight loss and strength loss of the steel bars*

(1) Weight loss of the steel bars

The nuts are removed and the prestress is released. After scraping off the mortar and the protective coat, it is found that there are not rusts on the steel bars' surface, and the steel bar is just lackluster, comparing with the unused steel bar. Therefore, it is thought that the steel bars don't lose weight.

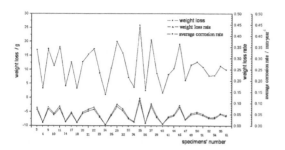

Figure 3. Graph of weight loss, weight loss ratio and average corrosion rate of the steel cylinders.

(2) Strength loss of the steel bars

The tensile test's results of the steel bars, which were applied with different grade of prestress, is given in Tables 6, 7, and 8. Results of the 55* and 54*

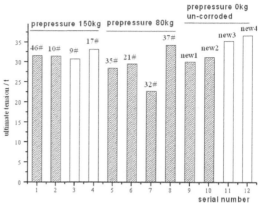

Figure 4. Strength loss comparison between the exposed steel cylinders and the new steel cylinders.

Table 6. Results of tensile test of the steel bars (pretension: 80 kg).

Specimens' number	Service time/ year	Yield load/ KN	Limit load/ KN	Extension/ mm	Fracture state	Mortar	Protective coat
37	17	1.73	2.78	8.95	Fractured at 0.73L **	Common mortar	None
30	17	1.80	2.83	8.70	Fractured at one end	Common mortar	Chlorosulfonated polyethylene coatings
33	17	1.90	2.77	10.08	Fractured at 0.98L**	Common mortar	Acroleic grout coat
49	17	2.52	2.98	7.35	Fractured at one end	Mixed with additive	None
55*	17	2.93	3.07	2.68			
Average	17	1.988	2.840	8.770			

Table 7. Results of tensile test of the steel bars (pretension: 150 kg).

Specimens' number	Service time/ year	Yield load/ KN	Limit load/ KN	Extension/ mm	Fracture state	Mortar	Protective coat
47	17	1.66	2.63	7.35	Fractured at 0.95L **	Common mortar	None
17	17	2.26	2.83	6.99	Fractured at 0.8L **	Common mortar	Chlorosulfonated Polyethylene coatings
26	17	2.56	3.00	6.79	Fractured at one end	Common mortar	Acroleic grout coat
23	17	1.67	2.55	6.82	Fractured at 0.7L **	Mixed with additive	None
54*	17	3.03	3.19	4.90	Fractured at 0.93L **	Acroleic mortar	None
Average	17	2.038	2.753	6.988			

101

Table 8. Results of tensile test of the new steel bars (pretension: 0 kg).

Specimens' number	Service time/ year	Yield load/ KN	Limit load/ KN	Extension/ mm	Fracture state	Mortar	Protective coat
1	0	3.24	3.74	6.4	Fractured at 0.93L*	None	None
2	0	2.94	3.53	5.47	Fractured at 0.93 L*	None	None
3	0	2.94	3.63	6.36	Fractured at 0.93 L*	None	None
Average	0	3.040	3.633	6.077			

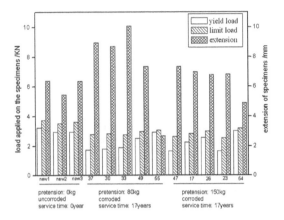

Figure 5. Histogram of fracture load and extension of the steel bars.

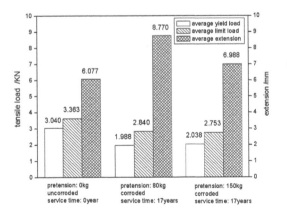

Figure 6. Histogram of average fracture load and average extension of the steel bars.

specimens are not used for calculating the average values. L is the length of the steel bars.

Fig. 5 and 6 are the histograms of Tables 6, 7 and 8.

4 ANALYSES OF TEST RESULTS

4.1 Analyses of corrosive conditions

(1) Corrosive conditions of stratum are usually divided into 4 grades, viz. very strong, strong, medium and weak. Table 4 shows that PH value of Zhucun colliery is 7.44, belongs to the medium grade, 6.5~8.5; CO_2 content of the groundwater of Zhucun colliery is 14.0 mg/L, nearly 2 times higher than that of the strong grade, 5 mg/L; SO_4+Cl content of the groundwater of Zhucun colliery is 29.58 mg/L, belongs to the weak grade, <100 mg/L. Therefore, according to the common principle of confirming corrosive grade, the corrosive grade of Zhucun colliery is the medium grade.

(2) Temperature is the catalytic factor of corrosion. The higher temperature is, the greater the corrosion rate is. Usually, relation between corrosion rate and temperature is exponential relationship. In this field test, the temperature of the test site nearly keeps invariable; therefore, temperature only had very little influence on the corrosion rate.

(3) Humidity can influence the corrosion greatly. Moisture in air is a necessary condition for both metal ionization and medium ionization. Besides participating in the basic process of corrosion, moisture also has influence on the other factors relating to corrosion. If the moisture content is too high, the diffusion and osmosis of oxygen will be obstructed, and the corrosion rate will be decreased. With the increase of the moisture content, medium resistivity will decrease, and the effect of oxygen concentration cell will increase to its peak when the humidity increases to 70~90%. If the medium humidity increase any more and is close to saturation, the effect of oxygen concentration cell will decrease, and the corrosion rate of bolts will be decreased.

(4) According to the assay results of the groundwater, the groundwater is neutral, totall acidity is on the high side, and the Cl−content is the highest among all anions. The salinity of soil usually is

80~1500 ppm. Kation in soil usually are potassium ion, sodium ion, magnesian ion and calcium ion. Anions in soil usually are $Cl-$, SO_4^{2-} and CO_3^{2-}. The higher the salinity of soil is, the higher the conductivity and the corrosion ability of soil are. Therefore, for this test, the corrosive effect of $Cl-$ was remarkable on the model bolts.

4.2 Corrosive state analyses of the exposed steel cylinder

Analyses of Fig. 2 and Table 4 give the following results.

(1) It is important to analyze the corroded state of the exposed steel cylinder, because the steel cylinders simulated badly constructed bolts.
(2) Pitting corrosion of each steel cylinder is non-uniform. All pitting corrosion state of all steel cylinders is different from each other. Pitting corrosion depth is 0.15~4.0 mm. Only the maximal corrosion depth can be used as the controlling parameter. Therefore, it is thought a steel cylinder of Φ 40 will lose 10% diameter and 19% section area after 17 years.
(3) Corrosion of each steel cylinder has certain distribution rule. Most pitting corrosion happened near the contact position of the steel cylinder with the drilled anchor-hole. Probably, the reason of this phenomenon is that the contact poison was more humid.
(4) Steel cylinders of No. 1, 2, 5, 6, 7, 8, 12, 13, 15, 16, 25, 27, 28, 31, 38, 43, 53, 60 and 62 were not corroded after one year. This shows that pitting corrosion was progressive. The pitting corrosion depth in each year was not same. Therefore, avenge corrosion rate should not be used to calculate the service life.

4.3 Weight loss, weight loss ratio and average corrosion rate of steel cylinder

Analyses of Fig. 3 and Table 5 give the following results.

(1) Even under same corrosive conditions, the weight loss of every steel cylinder is extremely non-uniform. The weight loss of each of 7 steel cylinders, 20.6% of all steel cylinders, is lower than 5 g. The weight loss of each of 9 steel cylinders, 26.5% of all steel cylinders, is higher than 15 g. The weight loss of other 18 steel cylinders, 52.9% of all steel cylinders, is 5~15 g. The probability distribution of the weight loss is approximately normal distribution.
(2) Weight loss ratio of the steel cylinder is the ratio of its lost weight to its original weight. The probability distribution of weight loss ratio is also approximately normal distribution. The highest weight loss ratio is 0.122%, 24.4 times of the lowest, 0.005%.
(3) Calculation formula of average corrosion rate is as follows,

$$V = \frac{365(W_1 - W_2)}{2\pi R(L+R)T\rho} \tag{1}$$

where, V is the corrosion rate of the steel cylinder in mm/year; W_1 is the original weight of the steel cylinder in g; W_2 is the weight of the steel cylinder in g; π is circumference ratio, 3.1415926; R is the outer radius of the steel cylinder in mm; L is the length of the steel cylinder in mm; T is the corrosion time of the steel cylinder in day; ρ is the density of the steel cylinder in g/mm^3.

Formula (1) is established on the concept of weight loss. Therefore, the tested average corrosion rate also has same distribution shape with weight loss ratio. Average corrosion rate is based on the postulate of uniform corrosion. The actual corrosion state is extremely non-uniform. Therefore, the calculated result with formula (1) is not safe. For one steel cylinder, weight loss ratio is also an average value. This is the main reason that the tested average corrosion rate also has same distribution shape with weight loss ratio. The greatest pitting corrosion depth and the corresponding corrosion rate should be most important.

4.4 Strength loss ratio of steel cylinder

Analyses of Fig. 4 give the following results.

(1) The limit tensile loads of 12 steel cylinders are dispersive. In these specimens, the fracture of 8 specimens happens near the screw thread or the welded seam. This shows that the screw thread reduces the strength of the specimens and the strength of the welded seams is not enough. This leads to the distortion of the data. (2) For No. 9 and No. 17 steel cylinders and the new No. 3 and No. 4 steel cylinders, the fracture happens in their middle position. This shows that the strength of these steel cylinders' welded seams is enough. The tested data of these specimens have higher degree of confidence. (3) After necessary error correction, the strength loss ratio of the tested steel cylinders are confirmed to be 7.55~13.26%. For safety, it is thought the strength loss ratio of the specimens, which had been in service for 17 years, should be 14%.

4.5 Fracture loads of the steel bars

Analyses of Tables 6, 7, 8 and Fig. 5, 6 give the following results. (1) The yield load, limit load and extension of the steel bars, which were protected by acrylic grout, are highest among all the steel bars. This shows that acrylic grout has the best protection effect. Other steel bars' yield loads, limit loads and extensions are nearly equal to each other. There are not rusts on all steel bars. This show it is not important whether there is a coat on the steel bars. This also shows the steel bars, except the steel bars that had been protected by acrylic grout, also can be analyzed with averaging their yield loads, limit loads and extensions. (2) According to Fig. 4 and 5, the average yield load of the new steel bars, which have not been applied with pre-stress and not been under the corrosion condition, are 52.9~49.2% higher that of the steel bars, which have been in service, been applied with pre-stress and been under the corrosive condition for 17 years. The average limit load of the former is 18.4%~22.2% higher than that of the latter. The average extension of the former is 30.7%~13.0% lower than that of the latter. This shows the yield load and limit load of the latter are remarkably reduced and the extension of the latter is remarkably improved. Apparently, this is not caused by the rust-eating. (3) The corrosive conditions of the steel bar applied with 80 kg pre-tension are same with those of the steel bars applied with 150 kg pre-tension. Therefore, the difference between the former and the latter should be relevant to the stress corrosion. The average yield load of the former is 2.3% lower than that of the latter, but the limit load of the former is 3.2% higher than the latter. It is usually think that material will be fractured by stress corrosion when the tensile stress achieves 70%~90% of the yield strength of the material. For this test, because the designed average tensile stress of the former and the latter was only 39.4% and 72.1% of their yield stress respectively, it is difficult to distinguish the difference of the stress corrosion effect between the former and the latter. (4)The strength reduction of the steel bars due to stress corrosion should exist but is unremarkable. However, the average limit load of the steel bars, which have been in service for 17 years, is 18.4%~22.2% lower than that of the new steel bars. Why is there so remarkable difference between these two kinds of steel bar? Both rust-eating and stress corrosion are not the reasons. One possible explanation is material's ageing. Another possible reason may be the improvement of material's property due to the improvement of smelting techniques. Anyway, this fact should be paid much attention. (5) The steel bars simulated the perfectly constructed bolts. Their appearance and weight didn't change after 17 years. They were also not rust-eaten. However, it is found that their average tensile strength is reduced largely in the tensile test. In addition, it is impossible that the actual construction of bolt achieve so perfection. The steel cylinders simulated the actually constructed bolts very well. After 17 years, they were rust-eaten in

different grades, and serious pitting corrosion even penetrated the wall of some steel cylinders. Their strength loss ratio is about 14%, diameter loss is about 10%, sectional area loss is about 19%. Therefore, the quality of construction has obvious influence on the endurance of bolt.

5 CONCLUSIONS

(A) Corrosive grade and conditions

(1) The corrosive grade of Zhucun colliery is the medium grade.
(2) The temperature of the test site nearly kept invariable; therefore, temperature only had little influence on the corrosion rate. The specimens were in the closed and humid conditions. Therefore, humidity influenced the corrosion of the specimens greatly. The corrosive effect of $Cl-$ was remarkable on the specimens.

(B) Corrosion of the steel cylinders

The steel cylinder simulated the bolt, which was constructed badly and under serious chemical and stress corrosion conditions.

(3) Pitting corrosion depth of each steel cylinder is different from the others. It is dangerous to evaluate the steel cylinders' corrosion with the average corrosion rate. It is better to evaluate the corrosion with the corrosion rate corresponding to the maximum pitting corrosion depth.
(4) The weight loss, weight loss ratio and average corrosion rate of the steel cylinders obey normal distribution. The weight loss ratio of all steel cylinders is extremely non-uniform. The highest is 24.4 times of the lowest.
(5) Under the medium grade of corrosive conditions, the strength loss ratio of the steel cylinders, which had been in service for 17 years, is about 14%, their diameter loss is about 10%, and their sectional area loss is about 19%. This means that the actual bolt, which is designed according to material's strength and the safety factor ≥ 1.14, will be fractured after 17 years in such corrosive conditions.

(C) Corrosion of the steel bars

The steel bar simulated the bolt, which was constructed well and under perfectly chemical corrosion and stress corrosion conditions.

(6) Acrylic grout has the best protection effect than other protective coats. In addition, if there is enough thick mortar around steel bar, it will be not important whether to coat on the steel bar additionally.
(7) Due to various factors, the average yield load and limit load of the steel bars, which had been in

service for 17 years, are 52.9~49.2% and 18.4~22.2% lower those of the steel bars respectively. This should be paid much attention.

REFERENCES

Rokhlin, S.L. et al. 1999. Effect of pitting corrosion on fatigue crack initiation and fatigue life. *Engineering Fracture Mechanics* 62 (4) : 425–44

Harlow, D.G. & Wei, R. 1995. Probability modeling for the growth of corrosion pits. In: Chang C. I., Sun C.T., ed. *Structural integrity in aging aircrafts*. ASME, 185–194

Bamforth P. 1996. Predicting the Risk of Reinforcement Corrosion in Marine Structures, Corrosion Prevention & Control

Amey, S.L. et al 1998. Predicting the Service Life of Concrete Marine Structures: An Envirpmental Methodology. *ACI Sturctural Journal*, March–April

Service-life Prediction, State-of-the-art report, ACI 365. R-00, Reported by ACI Committee 365, April 2000

Liu X.L. 2001. Basal Research on Durability of Structure Work[A]. In: Chen Z.Y. et al. ed. *Forum of Engineering Science and Technology: Safety and Durability of Construction Structures*. Beijing: Tsinghua University Press: 200~ 206. (in Chinese)

Wang Y.L. & Yao Y. 2001. Research and Application on Durability of Concrete in Importance Work. Beijing: China Architecture and Building Press (in Chinese)

Zhang M. 2001. Analysis on Safety and Durability of Chinese Railroad Tunnel Structures. In: Chen Z.Y. et al. ed. *Forum of Engineering Science and Technology: Safety and Durability of Construction Structures*. Beijing: Tsinghua University Press: 1~ 4. (in Chinese)

Yao Y. 2001 . Durability of Concrete Material – Research Evolve on Safety of Concrete in Importance Work. In: Chen Z.Y. et al. ed. *Forum of Engineering Science and Technology: Safety and Durability of Construction Structures*. Beijing: Tsinghua University Press, 266–273. (in Chinese)

Chen Z.Y. 2001. Research on Durability and Service Life of Concrete Structures. In: Chen Z.Y. et al. ed. *Forum of Engineering Science and Technology: Safety and Durability of Construction Structures*. Beijing: Tsinghua University Press, 17~ 24. (in Chinese)

Zeng X.M. et al. 2002. Discussion about "Time Bomb" for Bolt – an Answer to Professor Guo Yingzhong. *Chinese Journal of Rock Mechanics and Engineering*, 21(1): 143–147. (in Chinese)

Zhou S.F. 1998. Research on Durability of Concrete with Erode Surroundings in Underground Work. *Protective Engineering*, 6(1): 43–48.(in Chinese)

Zeng X.M. et al. 2004. Research on Safety and Durability of Bolt and Cable-supported Structures. *Chinese Journal of Rock Mechanics and Engineering*, 23(13): 2235–2242. (in Chinese)

Evaluating the ground pressures on the TBM and the lining for the Guadarrama base tunnel

Davor Simic
Polytechnical University of Madrid

ABSTRACT: The construction of base tunnels crossing mountain ranges is becoming increasingly common in the context of high speed railway lines. Such deep tunnels compared to the more usual shallower ones pose a new kind of problems, particularly in the area of ground squeezing phenomena, having great impact on the forces acting on the TBM machines and on the lining. This paper describes the geomechanical model of the rock mass along the new Guadarrama Tunnel for the high speed rail ling north of Madrid, for the purpose of evaluating ground pressures by means of numerical models. Rock mass deformability is assessed by means of different "in situ" tests such as dilatometer and seismic wave measurements, particularly in the faulted zones, more prone to squeezing behaviour. Three-dimensional numerical models by means of Finite Difference techniques have been run to evaluate ground forces in different locations of the alignement. This paper shows the advantages of having a reliable foreknowledge of the ground behaviour to limit the uncertainties during the tunnel construction.

1 INTRODUCTION

The every day more frequent construction of base tunnels through mountain ranges raises new problems that did not require analysis in tunnels that, due to their shallower depth and better quality of the rock mass, the ground deformability is not as critical. Accepting the inevitable fact that the improvement of communications through orographic barriers, and limiting ourselves to our country, recent works such as the Guadarrama tunnels, recently finished or the Pajares tunnel, just starting, will not be isolated examples in the context of underground works, making it necessary to study in detail important aspects of ground-tunnel interaction that until now had been avoided. Given the safety and performance requirements imposed for this type of construction it is usual the adoption of a tunnel boring machine for the excavation and lining with prefabricated segments, in which the convergence of the excavation is rigidly limited both by the shield and the lining ring. The study, mandatory for any tunnel, of loads induced by the rock is especially relevant due to the high level of pressure derived from the combination of a great overburden (which can attain several hundreds of metres) and the deformability of the rock mass (due either to the soft nature of rock matrix or the existence of faults in hard rocks). Furthermore, rheological phenomena become more important and, their integration in the constitutive laws of the material is still lacking of empirical confirmation, as it has not been compared with the instrumentation of real cases. However, the creep of the rock may represent a noticeable increase in lining requirements. This paper studies the influence of the rock deformability on the pressure generated both against the lining ring and the shield, which has great relevance in the construction feasibility of the tunnel, considering that the longitudinal thrust capacity of the tunnel boring machine (whose magnitude is currently limited by present technology to values less than 20,000 T) should be capable of overcoming the friction induced by the rock pressure, providing in addition the necessary pressure at the tunnel face and, hence, allowing the progress of the excavation of the machine.

2 INFLUENCE OF THE METHOD OF CONSTRUCTION ON THE GROUND PRESSURES

It is known that the excavation of a tunnel causes a stress relief at the face by virtue of which a deformational field arises in the medium whose vectors are aimed at the excavation. The magnitude of these convergences increase with the depth of the tunnel and deformability of the rock mass.

In deep tunnels it is unavoidable that, with the deformations an annulus of plastified rock will appear

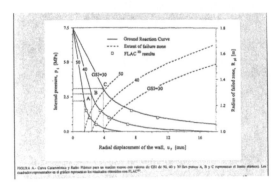

FIGURA A.- Curva Característica y Radio Plástico para un macizo rocoso con valores de GSI de 50, 40 y 30 (los puntos A, B y C representan el límite elástico). Los cuadrados representan en el gráfico representan los resultados obtenidos con FLAC³⁰.

Figure 1. Characteristic curve and Radius of failed zone around a cavity for three values of GSI (Serrano, 2004).

Figure 2. Rock TBM for the construction of the Gotard base tunnel.

around the excavation, in which the stress state reaches the rock failure envelope and displacements increase progressively. Figure 1 (Serrano, 2004) reflects this behaviour through the application of an analytical model of characteristic curves to a rock whose strength is based on the Hoek-Brown failure criterion.

In the event that such displacements should not be contained by the structural elements of the tunnel lining, the plastic annulus progressively increases its thickness, thereby increasing further the displacements in an unstable process that leads to the collapse of the cavity. The construction methods of the tunnel to counterbalance these displacements can adopt two possible strategies (Barla, 2001): active or passive approach. In the active approach, the objective is to avoid the convergences of the rock through a strong support and lining system capable of absorbing high thrusts derived from the short relaxation allowed to the characteristic curve of ground before being counterbalanced by the reaction of structural elements.

On the other hand, in the passive approach, the tunnel support system admits important deflections in a controlled manner in order to reduce the demands imposed by the ground. Given that, in deformable terrains, the stiffness of ordinary support elements (shotcrete or steel in the case of trusses) is not compatible with the range of convergences to be allowed in order to achieve a notable reduction of pressures, in practice additional, measures have been adopted such as over excavation of a rock thickness equal to the expected convergence, the adoption of longitudinal compressive joints in the shotcrete support or TH profile sliding trusses. In the case of a completely mechanized excavation (tunnel boring machine) as for example in figure 2, in which the size of the excavated area is fixed, this approach stumbles with the difficulty of predicting the magnitude and distribution of convergences, therefore it is not easy to select

support measures during the excavation process. To add more complexity to the problem, the rate of progress has considerable influence on the ground deflections, so it is likely that the tunnel support should be increased in successive stages in the light of the readings of the instrumentation.

In any case it is considered good practice that, regardless of the approach being adopted (active or passive), the provisional support and lining system, shall have a more or less circular shape with continuity around the excavation perimeter to provide the necessary structural strength with controlled deflections, immediately behind the face, which is hardly compatible with a construction procedure based on the partial excavation of the section. In the case of standard cross sections of single rail railway tunnels, this involves the use of tunnel boring machines with a cutting head that allows the circular excavation of the entire section and the installation near the face of the provisory support or final lining. In addition, in the case of deformable ground, the use of an open TBM as shown in figure 2, has the disadvantage that the unavoidable collapses that are associated to the convergences hinder both the action of "grippers" and the placing of the temporary support; problems that can be overcome if a "shielded" TBM is used.

In the case of a shield-based construction process and lining of segments (see figure 3), the cutterhead produces a circular section excavated of a diameter greater by several centimetres to external diameter of the shield. At the same time, the shield presents a certain "conicity", hence the external diameter of the tail is smaller than the front body. Finally, given that the lining ring is erected on the inside of the shield, a ring-shaped space is left between the extrados of the segmental ring and the excavation surface. Such space is filled by different procedures (gravel, grout injection, etc.) and gradually assumes load as the

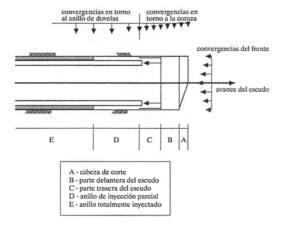

A - cabeza de corte
B - parte delantera del escudo
C - parte trasera del escudo
D - anillo de inyección parcial
E - anillo totalmente inyectado

Figure 3. Distribution of ground convergences in the construction process with a tunnel boring machine and lining of segments.

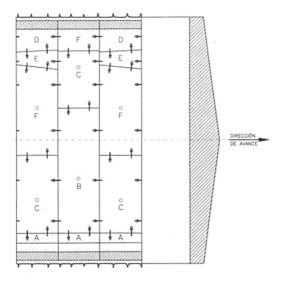

Figure 4. Longitudinal schematic section of the lining ring installed by a shielded TBM.

tunnel boring machine advances; therefore there is a section around the shield (see figure 4) in which the ground may converge more or less freely, although in the final situation the concrete segment ring will rigidly ground deflection through hardened or compact in fill material. This method avoids the aforementioned risks for the open TBM, although the shield is more sensitive to the risk of trapping by convergences of the ground.

3 RISK OF TBM BEING BLOCKED IN THE "LA UMBRIA" FAULT

3.1 Description of the works of the Guadarrama tunnels

This work is the essential strech of the New Railway Access to the North and Northeast of Spain, for the exploitation of high speed trains. The need to overcome the topographical accident of the Madrid mountain range and in order to preserve important natural protected spaces, a tunnel of 28,377 m was required.

For these lengths, the safety operating conditions require the adoption of a cross section consisting of two one way tunnels intercommunicated through evacuation galleries every 250 m (see figure 4). In order to design the internal section of the tunnels, it has been considered the requirements of trains and aerodynamic conditions, adopting a circular section with an internal diameter of 8.50 m. In addition, other underground works are being built such as underground transformation centres every 2,250 m and an emergency chamber. The ground to be excavated by the tunnels mainly consists of glandular gneisses, with interspersed granite in Soto del Real, the Eresma Valley, Paular and Lozoya river. The main expected geological-geotechnical risks are:

- Faulted area of La Umbria, where a Cretaceous basin appears near the tunnel level with sands and limestone between the faults, and a thick strip of crushed rock.
- Deep overburden in Peñalara, around 1,000 m.

With regards to hydrogeological conditions, the gneisses and granites offer very low permeability, except in fault areas, where maximum inflow values were lower than 100 l/min.

From the geotechnical investigation of the rock, it can be stated that the unconfined compression strength presents widely scattered values, with an average in gneisses of 70 MPa and granites of 76 MPa. The point load tests have registered an average value of $I_s(50)$ around 8,35 MPa both for gneisses and granites. The average values of RQD are around 64% in the gneiss, 57% in granite and 7% in faults. The average quartz contents amounts 31.8% in gneiss and 27.8% in granite.

The deformation moduli have been measured with dilatometer tests, and estimated from complete sonic wave records in boreholes, as well as sample tests in laboratory. The latter presented values of 20 to 50 GPa, while the dilatometer ones resulted in 10 to 20 GPa.

Given the characteristics of materials to be excavated, with a net predominance of good quality rock, one is lead to think that the type of machine required will be an open TBM. However, the existence of fault areas, specifically around the La Umbría Valley, meant the need to consider the possibility of using a

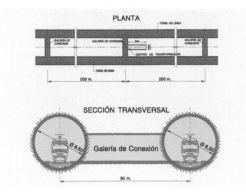

Figure 5. Standard scheme of Guadarrama tunnels.

Figure 6. Articulated telescopic study (double shield) for perforating tunnels.

double shield, machine capable of also excavating in unstable ground.

Typical rock TBM allows perforating tunnels of great length with very high performance in competent rock, but it slows down in fault areas due to the difficulty of the ground to sustain the reaction to the gripper's thrust.

Double shields, on the other hand, participate in the characteristics of open TBM and single shielded TBM. Open machines obtain the necessary reaction for the advance by means of diametrically expandable grippers thrusting against the tunnel walls while in the shielded TBM the segment ring is erected in the tail and the reaction for advance is obtained by the longitudinal jacking against the segment lining already in place. Double shields have grippers built into the rear shield, having thus dual thrust capacity: either against the rear shield (which allows to install the lining simultaneously with the excavation) or against the already in place lining, when the rock conditions do not allow effective and adequate gripping.

In this work, the existence of important fault systems, both in the Lozoya Valley and, especially, in the Umbria Valley, the aforementioned considerations favoured the selection of a double shield instead of an open shield.

Figure 5 shows the machine, which is operated by electricity with a power of 4,200 kW and torque of 4,000 N.m. The cutterhead, with a diameter of 9.51 m is equipped with 61 disks of 17 inches diameter. The machine has the following elements to correct the various eventualities:

• Copy cutters that allow excavating an overcutting that caters for the convergence of the ground, reducing pressures on the shield and segments.
• Possibility of injecting bentonite through the tail skin.
• Possibility of performing investigation boreholes and treatment of ground ahead of the face.

The segmental lining (see figure 6) consists of 6 segment (plus a key segment) of reinforced concrete 32 cm thick. Waterproofing of the tunnel is achieved with two barriers, the first consisting of ethylene-propylene gaskets fitted in existing grooves near the extrados and a secondary one consisting of the grouting of annulus between the ground and the segment extrados.

3.2 *Geomechanical investigation of the rock mass at the fault zone*

Practically equidistant with regards to both portals the alignement intercepts the faulted area of Valdesqui-Umbria-Lozoya, which presents great thickness towards Southeast wedging towards Northeast, where a set of raised and sunken geological blocks are bounded by a large number of shear zones where the gneissic rock is totally fractured and even of sandy, appearance. Some strips of sand and sandstone corresponding to the Cretaceous (Albian) dragged by the tectonic forces are also detected here. The geological plant of figure 7, at the tunnel level, shows an interpretation based on the investigations made, where it is possible to see how the original design alignement was affected around Km 33 + 300 by a wide band of poorly cemented and saturated cretaceous sands. Furthermore, additional boreholes showed that this band disappeared towards Northeast, as shown in the transversal cross section to the alignement in figure 8. In order to avoid the risk of excavating in sands under water pressure, it was decided to move the alignement by 230 m towards the Northeast in order to reduce the probability of intersecting cretaceous sand as much as possible, which in this area, is located below the tunnel as can be seen in the longitudinal profile of the new alignement in figure 9.

In order to characterize geotechnically the materials of the fault (glandular and blastomylonitic gneisses,

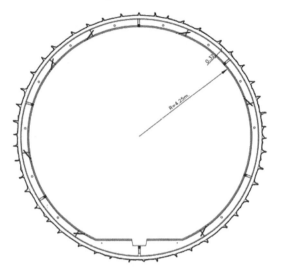

Figure 7.   Guadarrama tunnel. Lining ring.

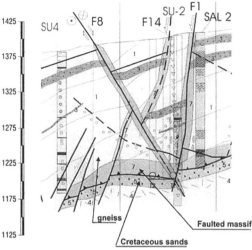

Figure 9.   Geological cross section profile in the faulted area of La Umbria.

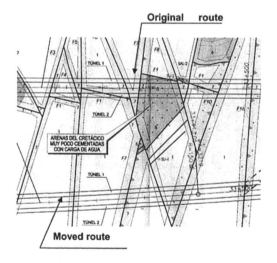

Figure 8.   Faulted area of La Umbria. Geological plant at tunnel level.

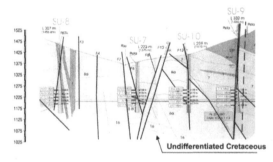

Figure 10.   New alignement in the faulted area of La Umbria. Geological-geotechnical longitudinal section.

sometimes highly fractured and sandy), a complementary survey of deep borings was performed with sample recovery for laboratory tests, execution of dilatometer tests, geophysical testing with complete sonic wave records, oriented image and resistivity/spectometry of natural gamma radiation. In addition, a seismic tomography campaign was carried out through blasts at different depths in a borehole and wave detection in neighbouring boreholes, with hydrophones located at every 10 m in depth. The geophysical testing has allowed obtaining continuous records of the ground properties that have been very useful in order to establish the range of rock quality variations. Of the multiple correlations performed figure 10 shows as an example the relation between the RMR index obtained from the core recovered and the dynamic elasticity modulus.

From the analysis of data, it has been possible to identify and characterize three differentiated geotechnical units in terms of their stress-strain behaviour:

• Fresh rock (Grade I) with 4 fractures by metre and with RMR greater than 60. Figure 11 summarizes the results of dynamic and static moduli, respectively obtained from the sonic records and dilatometer tests.
• Highly fractured and somewhat weathered rock (Grade II-III), with variable RMR between 35 and 60. Figure 12 summarizes the values of the static and dynamic moduli.

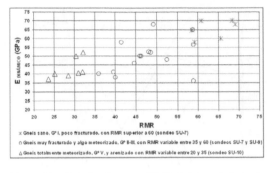

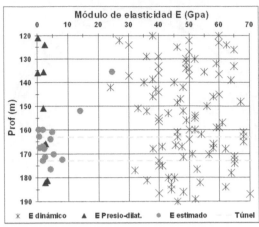

Figure 11. Relation between RMR index and dynamic elasticity modulus.

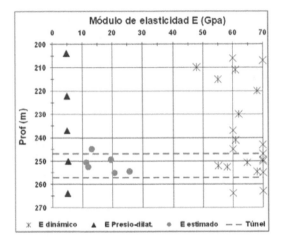

Figure 13. Fault of La Umbria. Static and dynamic elastic moduli at the tunnel for grade II – III rock.

Table 1. Stress-strain parameters of the rock mass at La Umbria fault.

Material	Unconfined compressive strength (MPa)	Rock mass modulus (MPa)
Fresh gneiss (Grade I), somewhat fractured, with RMR >60	20–75 42	5000
Highly fractured and weathered gneiss (RMR 35–60)	1.5–72 22	1900–3200
Fault zone, totally sandy	0.5–7.5 4	300–700

Figure 12. Fault of La Umbria. Static and dynamic elastic moduli at the tunnel for Grade I rock.

• Completely weathered (Grade V) and sandy rock. Figure 13 summarizes the values of the static and dynamic moduli.

Also, the elastic modulus can be estimated as a function of the RMR using the expression (Hoek et al, 2002):

$$E = \sqrt{\frac{\sigma_{ci}}{100}} \cdot 10^{\frac{GSI \cdot 10}{40}} \quad GPa$$

σ_{ci} = Resistance to unconfined compression.
GSI = Geological strength index.

Table 1 attached summarizes the main parameters of the massif in the area of the fault.

As a conclusion of the results, it can be seen that the moduli obtained from the pressuremeters are somewhat less than estimations based on RMR. Furthermore, the results of the tests show an average relation E_{DIN}/E_{EST} of 31.8.

3.3 Evaluation of the ground pressures against the shield of the TBM

In order to analyze the risk of trapping of the TBM a three-dimensional model of finite differences was prepared to simulate the excavation process of the tunnel boring machine, considering both overexcavations and the effect of grouting the annulus in the extrados of the segments, as indicated in figure 15.

In a first analysis, we studied the behaviour of a single gallery without considering the excavation of the adjacent, given that it is the initial situation in the construction phase in which the two tunnel boring machines are very distant from each other. In order to analyze the most unfavourable situation, we have considered an overburden of the 300 m, which is the maximum depth of the tunnel through the fault area of La Umbria.

Figure 16 shows the FLAC^{3D} mesh block of finite differences and figure 17 shows the displacement

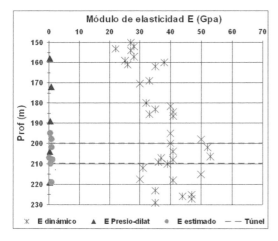

Figure 14. Fault of La Umbria. Static and dynamic elastic moduli at the tunnel for Grade V rock.

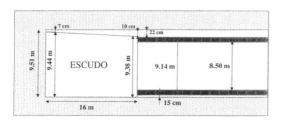

Figure 15. Geometry of overexcavations to be considered in the three dimensional model of finite differences FLAC^{3D}.

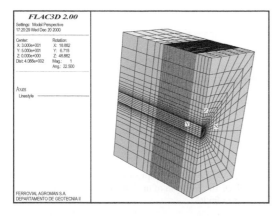

Figure 16. Diagram block of elements of the FLAC^{3D} 3D model.

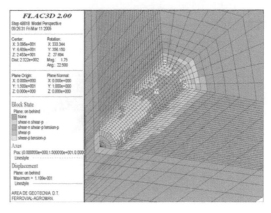

Figure 17. Displacements and plasticity indicators.

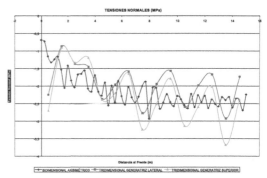

Figure 18. Normal stresses against the shield.

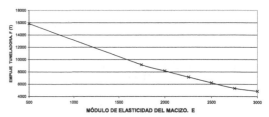

Figure 19. Analysis of sensitivity of the friction resistance of the shield with the elastic modulus of the ground.

vectors and plasticity indicators of the ground elements. Convergences of the rock induce normal stresses on the shield that are represented in figure 18, where the values of pressure corresponding to the extremes of both diameters have been superimposed with the pressures obtained from an axisymmetrical model. The integration of these normal stresses multiplied by the friction coefficient, through all the lateral surface of the shield shell allows assessing the resistance to the advancement of the TBM imposed by the convergence of the massif, which must be counteracted by the longitudinal thrust provided by the jacks.

We have analyzed the sensitivity of the calculated thrust to the rock modulus, as shown in figure 19. For

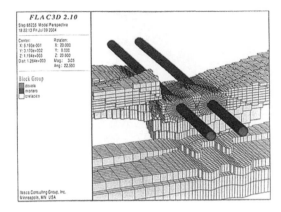

Figure 20. Partial view of the 3D element mesh of the FLAC^{3D} model showing the tunnels and the Grade V material.

the maximum thrust capacity of the TBM, which can be fixed at 12,000 t, it can be seen that the convergences of the ground can block the tunnel TBM for an average elasticity modulus of less than 1,300 MPa that, in the case of the La Umbria fault, correspond to Grade V materials of highly fractured and weathered gneiss or Cretaceous sands.

This preliminar assessment has allowed to confirm the importance of avoiding the more altered and sandy rock sections where the risk of blocking the TBM is high.

However, we should state that this assessment is very much on the safe side, as it assumes that the tunnel overburden is 300 m when the most unfavourable geotechnical conditions are presented and, on the other hand, it supposes an homogenous rock mass with the more unfavourable deformable ground characteristics.

4 DETAILED RHEOLOGICAL MODEL OF THE FAULT INTERSECTION

Once the geological-structural conditions of the fault have been identified, we performed a 3D model considering the spatial distribution of the various materials. Figure 20 shows a partial view of the mesh where the two galleries and the more weathered rock are shown. We used a constitutive equation that includes a creep law to consider the rheology of materials (see figure 21), which allows assessing the long term forces against the lining. Figure 22 shows the friction resistances due to the convergence of the ground around the shield, where it is shown that the maximum value is around 15,000 t. Finally figures 22 and 23 show, respectively, the maximum values of the compression stresses in the lining for the first tunnel and after the excavation of the second. It can be seen that the maximum compression of the lining is

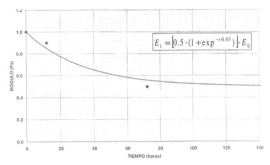

$$E_i = \left[0.5 \cdot (1 + \exp^{-t \cdot 0.03}) \right] \cdot E_0$$

Figure 21. Variation of the elastic modulus over time.

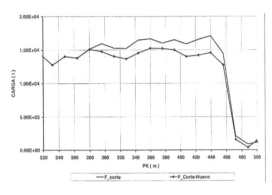

Figure 22. Total friction resistance vs position of the tunnel boring machine in the La Umbria fault.

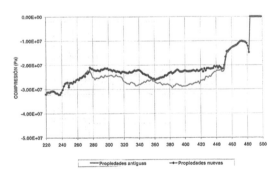

Figure 23. Maximum compression stresses of the lining in tunnel 1 vs, the situation of the tunnel boring machine in the La Umbria fault.

32 MPa and corresponds to the section in which the tunnel traverses the sandy cretaceous material, while in the remaining of the fault area, the maximum stresses are around 22.8 MPa.

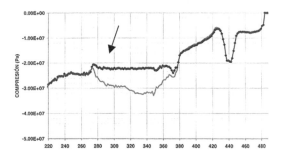

Figure 24. Maximum compression stresses of the segment in tunnel 1 after the 2nd tunnel was bored vs situation of the tunnel boring machine in the La Umbria fault.

5 PERFORMANCE DURING CONSTRUCTION

Towards the middle of the faulted area the TBM was blocked. Until that moment, the maximum thrust was between 2000 and 3000 t and the maximum torque 400 t. m, with an advance speed of 50 mm/min, and the TBM was operated with retracted grippers, jacking against the lining. The blocking was due to loose fragments of rock and sand material against the cutterhead. To consolidate the tunnel face chemical foams were grouted through short drills performed with the dedicated rig in the TBM. Also hand removal of loose material behind the cutterhead and the temporary closing of a 50% of the openings for the entrance of the muck was needed before the cutterhead began to rotate.

Except for this blocking event, the mobilized thrust of the TBM was constantly below the capacity of the jacks.

6 CONCLUSIONS

The importance to carry accurate evaluations of the ground pressure has been shown in a particular deep tunnel. To carry such analysis it has been necessary to obtain realistic parameters of the stress-strain behaviour of the rock mass and to implement three-dimensional models to evaluate the stress field due to the tunnel excavation and lining, taking into account the particular features of TBM construction.

The performance of the machine has confirmed the prognosis carried out, showing the reliability of the approach.

ACKNOWLEDGEMENTS

The author wishes to thank ADIF (Administrador de Infraestructuras Ferroviarias) for the kind permission to publish this paper.

REFERENCES

Barla, G. (2001). Tunnelling under squeezing rock conditions. Innsbruck.

Hoek, E.; Carranza-Torres, C.; Corkum, B, (2002). Hoek-Brown failure criterion-2002 edition. Proc. North American Rock Mechanics Society. Toronto.

Serrano, A (2004. Teoría de la plasticidad y rotura frágil. Jornada Técnica "tratamiento de Túneles en Roca". Sociedad Española de Mecánica de Rocas. CEDEX. Madrid.

Simic, D. (2005). Revestimiento de dovelas. Inyecciones de trasdós. Túnel de Guadarrama. Entorno Gráfico. Madrid.

Underground Works under Special Conditions – Romana, Perucho, Olalla (eds)
© 2007 Taylor & Francis Group, London, ISBN 978-0-415-45028-7

Crossing of fault zones in the MFS Faido by using the observational method

R. Stadelmann, M. Rehbock-Sander & M. Rausch

Amberg Engineering Ltd., Member of the Engineering Joint Venture Gotthard Base Tunnel South, Regensdorf-Watt, Switzerland

ABSTRACT: The paper deals with the successful crossing of severe fault zones in the multifunctional station Faido (MFS Faido), part of the Gotthard Base Tunnel (GBT), located in Switzerland. During the excavation work of the MFS Faido large, not prognosticated fault zones were found. Their impact on the layout and the tunnel support is described. Therefore the excavation procedure and the support system were specified over the combination of geotechnical computations and the observational method.

1   PROJECT OVERVIEW AND GEOLOGY

1.1   *The Gotthard Base Tunnel*

The Gotthard Base Tunnel's length of 57 km makes it the world's longest railway tunnel. It will be a high-speed rail link for passenger and heavy freight trains passing through the Alps between northern and southern Europe at top speeds of 200 to 250 km/h. The initial preparatory work for the GBT started in 1996; the tunnel is scheduled to be opened in the year 2015. The GBT consists of two parallel single-track tunnel tubes that are connected by cross-cuts every 312 m. To shorten the construction time and also for ventilation purposes, the GBT is divided into five construction sections (Erstfeld, Amsteg, Sedrun, Faido and Bodio). The tunnel is being driven simultaneously from both portals (Erstfeld in the north, Bodio in the south) and from three intermediate points of attack (access tunnels at Amsteg and Faido and a shaft at Sedrun). The Sedrun und Faido sections each possess a multifunctional station (MFS) (Fig. 1).

These stations contain four trumpet-shaped tunnel branch-off structures including two tunnel crossovers that allow trains to switch to the other tube, a ventilation tunnel system and emergency train stops that are linked with a side gallery. In an emergency occurring during the operating phase, the passengers can be evacuated through these side galleries. In addition, the MFS Faido consists of the cross cavern for housing the railway infrastructure equipment, the cross-cuts required for construction operations, the cavern for handling logistics in the construction phase.

Altogether there are about 6 km of tunnels and galleries with cross-sectional areas ranging from 40 m2 to more than 300 m2 in the MFS Faido. Its overburden ranges from 1,200 m to 1,600 m. The MFS Faido lies at the end of an access tunnel within the construction lot Faido and will be entirely excavated by drill-and-blast.

1.2   *Geology*

The MFS Faido was planned to be in favourable rock according to the geological prediction. Based on this forecast the entire MFS Faido lies in granitic Leventina gneiss that is favourable for construction. A few relevant but rather small fault zones were predicted, but these were not expected to have a significant effect on construction. After the 25th round had been blasted in the crown of the cross-cavern on April 11, 2002, a totally unexpected downfall occurred. As heading continued, faulty gneisses of low strength were encountered and rock started to loosen and fall at the face. This prompted intensive seismic and other investigatory operations. The results revealed a massive fault zone in the MFS area (Fig. 1)

2   THE IMPACT OF CROSSING FAULT ZONES IN THE MFS FAIDO

For safety and cost reasons, it was concluded that it would not be feasible to construct the branch-off structures in this geology. An intensive investigation programme was carried out to define the best place for the huge cross-over caverns with cross sections up to 300 m^2 and with an overburden of nearly 1,500 m. As a result of these investigations it was possible to adapt the layout of the multifunctional station with

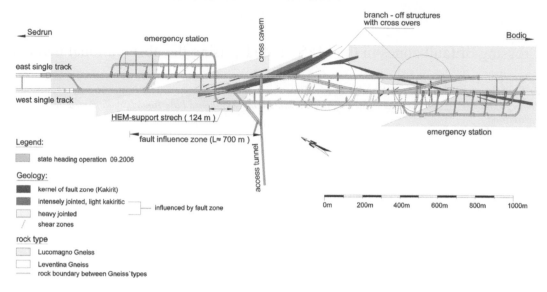

Figure 1.   Layout of the MFS Faido and revealed fault zones.

the aim of placing the large caverns in more favourable rock conditions in the southern part of the MFS. To date, all cross-over caverns have been successfully excavated.

Besides the layout of the MFS and the station's critical cross-sections, the geology encountered also made it necessary to carry out a critical review of excavation support means to be applied in the relevant cross-sections. Especially in the single-track tunnel west/north at the edge of the fault zone and in the transition zone between Leventina and Lucomagno gneiss (i.e. in the immediate vicinity of the fault zone), the initial support with pattern bolting and shotcrete was not appropriate to the new geotechnical conditions. The geological conditions encountered were interpreted as a combination of loosening and squeezing behaviour.

The initial support did neither provide the required stability nor the safety for the workforce. In order to ascertain the safety of the workforce this hazard could effectively be countered solely by means of a rigid support installed immediately behind the face. Because the layers and folds change from one meter to the next, a huge drilling campaign would have been required to achieve a precise spatial picture of the geology. And with only spotty knowledge of the three-dimensional structure of the geology, the excavation support requirements cannot be calculated reliably. The actual behaviour of the rock can only be detected by carrying out intensive measurements and adopting them respectively during heading operations. So the decision on excavation support can be made only on the basis of experience and close observation.

Based on the interpretation of the available convergence and extensometer measurements, a rigid support concept was discussed and developed. It was decided to carry out an in situ trial in the single-track tunnel west/north using extra-heavy steel supports backfilled with 40 cm cast in-situ concrete (double T-profile with HEM). Within the first 20 m using the HEM support type no problems were developed but the geological conditions were getting even worse.

The unexpectedly frangible or even cohesionless rock material made it impossible to leave enough free space between the heading face and the installation of the steel arches during the deformation phase. This meant that the arches had to be installed earlier and withstand greater loads. This produced radial convergences of more than 70 cm and in some cases fracturing of the HEM-profile beams. The rock bolting had to be increased to over 300 m of heavy steel bolts per running meter of tunnel. Under the enormous rock pressure, the steel arches began to yield. At some points, the entire support system was severed. The planned reserve space allowed for supplementary measures was therefore completely used up by the deformations. Drastic clearance problems arose within larger ranges.

For this reason the heading operations in the western tube towards North was finally stopped because the deformations in the back-up zone had reached excessive levels. The steel arches in these areas were severed completely, and safe working could no longer be ensured. The HEM stretch had to be rebuilt, because underbreak had been suffered over the entire length of about 124 m and it also affected the invert.

Table 1. Geotechnical parameters of rock types.

Geotechnical parameter		Rock type						
		A	F	G	H	J	D*	F*
		Leventina Gneiss	Schist	Phyllite	Kakirit	Kakirit	Kakiritic Gneiss	Kakiritic Schist
Compressive strength	[MPa]	12.03	1.44	0.69	0.48	0.92	1.50	0.72
Cohesion	[MPa]	3.0	0.4	0.2	0.15	0.3	0.4	0.2
Internal angle of friction	[°]	37	32	30	26	24	34	32
Young's Modulus	[GPa]	40	7.5	5	4	2	15	6

The HEM arches were dismantled and, following additional excavation, replaced by new supports (heavy flexible arches TH 44). The support system was changed to a more flexible steel construction in combination with shotcrete and anchors with convergence slots. This method is now being used successfully for more than 2 years.

3 GEOTECHNICAL COMPUTATIONS

In the construction project different rock types were defined which classify geological units and geotechnical parameter (Table 1). The predicted Leventina Gneiss is classified by rock type A. The severe fault zone and the sections influenced by it are classified by different rock types.

Firstly the fault zone in the MFS Faido was classified as kakiritic gneiss by rock type D*. Based on the growing knowledge of the Geology depending on the heading, the rock mass was determined to be kakiritic schist (rock type F*). In the single-track tunnel west/north two main factors have an unfavourable impact from the geological point of view. One factor is the relevant fault zone; the other is the folded change from Leventina Gneiss to Lucomagno Gneiss. According to the challenging geological conditions the rock pressure around the cavity is very high in vertical as well as in horizontal directions. Due to the fault dipping of 60° the rock pressure is asymmetric.

In order to understand the failings in the behaviour parameter studies were carried out with different methods. With the occurrence of the unexpectedly squeezing ground behaviour the original computations and dimensioning based on rock parameters and the characteristic line curve were carefully analysed and verified with the help of the Geological Strength Index (GSI) and by back analysis.

3.1 Back analysis HEM support – characteristic line method

At the time where the HEM support was developed hardly any knowledge about the rock mass and its

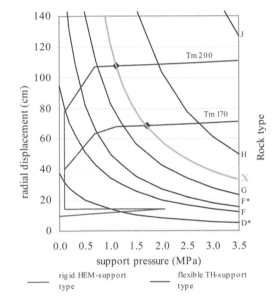

Figure 2. Characteristic lines.

behaviour of the fault zone existed. It should be noted that the support strength of the HEM arch system would have been sufficient if the bearing behaviour of the rock had corresponded to the material properties of the Rock Type D* (kakiritic gneiss) that were predicted.

Based on this information a back analysis of the HEM-Support type could be carried out using the characteristic line method. The characteristic line of the HEM support was reconstructed by the measured deformation and the installed support pressure as shown in Fig. 2.

It was assumed that the support pressure had dropped to a value of 0.1 MPa after overstressing. The radial displacement increased as long as the reconstruction type was not installed. The installation and the performance of the reconstruction type are shown by the blue coloured line. Depending on the measured deformation and the time when the deformation process

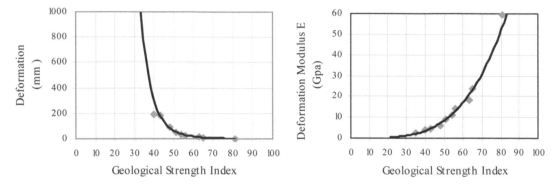

Figure 3. Relationship between the estimated GSI-value and the measured radial deformation; and the relationship between the estimated GSI-value and the analysed E-Modulus.

could be stopped the characteristic line curve (Rock type X) was assessed. According to the back analysis it was possible to estimate the bandwidth of the rock properties encountered (hatched area). This led to the choice of a flexible support system for further heading, which in the worst case would be capable of coping with the worst rock Type H (cohesionless zonal kakirite).

Experience of deep tunnel construction indicates that rock pressure decreases as the amount of rock deformation increases. The amount of constructional resistance needing to be applied can therefore be substantially reduced if a certain amount of deformation is accepted. The characteristic line method shows that the rock pressure in fault zones can only be stabilized if enough deformations are accepted. Otherwise very high support pressures were necessary, which would lead to uneconomic excavation and to an enormous thick lining. This resulted in a rock support concept where high degree of deformation is possible at the beginning but also high level of outward resistance is provided at the end.

3.2 Geological Strength Index method

The Geological Strength Index provides a system for estimation rock mass behaviour, especially the reduction in rock mass strength for different geological conditions. The aim of using the GSI-Method was to confirm the rock mass parameters which were assigned to the rock types. The method evaluates quantitative descriptions of rock mass properties as rock type, mineralogical composition, texture, schistosity, bedding, joints, faults, persistence etc. Due to the first excavation using the HEM-support type, a lot of geological information was available.

The used estimation of the GSI is based on the Rock mass rating system (RMR). Depending on the different input data the estimated GSI ranges between 35 and 81. The GSI of 81 is assigned to very good quality rock

mass and the GSI of 35 to poor quality rock mass. Based on the analysis of a number of case histories, which involved dam foundations for which the deformation modules were evaluated by back analysis of measured deformations. Figure 3 shows the relationship between the estimated GSI-value and the measured radial deformation; and the relationship between the estimated GSI-value and the analysed E-Modulus. In comparison to the rock type concept of the Gotthard Base Tunnel the estimation of the rock mass parameter with the GSI-Method leads to higher shear parameter and radial deformation based on a smaller young's modulus. Therefore the shear parameters are overestimated and the radial deformations are underestimated. In conclusion,

– the estimated parameter give the order of magnitude and need to be verified with numerical analysis,
– the estimated E-Modulus is similar to the E-Modulus of the defined rock types,
– the rock mass parameter of the rock types cover the sufficient bandwidth,
– the relationship between stiffness, strength and deformation is not applicable for the rock mass behaviour in the MFS Faido.

3.3 3D-analysis

From the rock mass parameter after the rock type concept deformation and plastic zones around the tunnel can be estimated using characteristic line and numerical analysis. Conversely, the effective rock mass parameter can be recalculated using the measured deformations. Other than that mentioned 3D-analysis were also conducted for the fault zone section. The design gives results for the deformation process of the rock mass considering the particular distance to the tunnel face. In addition, the 3D-analysis reassesses the geotechnical parameter. The assumptions of the 2D-analysis can be verified by the 3D-analysis. The 3D-analysis was conducted for

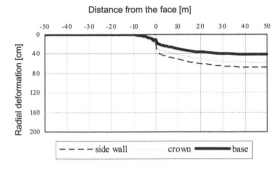

Figure 4. Radial deformation depending on the distance to the tunnel face for rock type F.

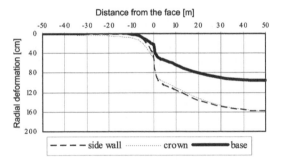

Figure 5. Radial deformation depending on the distance to the tunnel face for rock type F*.

a tunnel with a diameter of 12 m and an overburden of 1500 m using the rock type F*. The lateral pressure coefficient (relationship between vertical and horizontal stress) was assumed to be 0.5 or 1.0. If the in situ stress field were hydrostatic the radial displacements around the cavity would be very similar in on the roof, the side walls and on the invert. On the contrary the radial displacement would be asymmetric if the in-situ stress field were asymmetric.

Due to the sub-vertical dipping fault zone the radial displacement develop asymmetrical. Figures 4 and 5 show the radial displacement depending on the distance to the tunnel face for rock type F and F*.

The analysis leads to the following conclusion:

– the 3D-analysis verifies the 2D-analysis.
– 80% of the overall displacement need to be developed in order to change the support principal from flexible to rigid and additional shotcrete layers can be applied. This implies a minimum distance of 30 m for the deformable section. Otherwise the shotcrete will be overstrained.
– 90% of the overall displacement need to be developed before the last shotcrete layers can be applied.

3.4 Back analysis to estimate the rock loads for the inner lining

On the basis of the above described computational methods the new support type for fault zones in the MFS Faido were developed and successfully assembled. After the successful construction of the tunnel with the temporary support element the permanent lining had to be designed. In order to estimate the rock loads for the permanent lining numerical calculations were conducted. Therefore the construction, the installed support elements and the measured deformation need to be taken into account while conducting a Back-Analysis. This leads to different rock loads and their distribution on the lining which needs to be considered in the design.

The back analysis was carried out with a 2D-finite difference program including parametric studies (geotechnical parameter, lateral pressure coefficient). The constitutive law for the rock was based on the Mohr-Coulomb failure criterion. The simulated rock mass was divided into 5 sections. The fault zone was simulated with a width of 20 m and the influenced zone on both sides of the fault zone was 50 m. The excavations of three parallel situated tunnels were simulated depending on their heading process and construction method. The in-situ stress state depends on an overburden of 1400 m. The measured deformations in the MFS Faido show that the primary vertical stress is higher than the horizontal stress. This is considered with a lateral pressure coefficient of 0.5 and 1.0. The numerical approach includes the following steps:

– The support pressure is continually reduced and thus the characteristic line of the rock mass estimated.
– The comparison between the measured deformation of all measuring points and the calculated ones show whether the rock mass parameter and behaviour are evaluated correctly.
– Following this the characteristic line curves of the support are assigned for the location and in dependence on the particular measuring points.
– The rock load is assigned for every measuring point on the basis of the characteristic line method.

4 OBSERVATIONAL METHOD

In the case of the MFS Faido, where ground and structural behaviour could not be predicted with sufficient reliability by means of previously performed site investigations, structural analyses and comparable experience, the design have been carried out with the aid of the observational method. When using the observational method the information obtained on the geotechnical properties and the behaviour of the ground and the structure during construction have to be incorporated in the process of design and execution.

Within the observational method the following conditions are taken into account.

To monitor the behaviour of the ground and the structure control values corresponding alert and alarm values were fixed for the acceptable and critical range of behaviour.

Measures were planned and prepared together with additional safety measures, by which an inadequate behaviour of the structure could be dealt with or structural failure could be excluded.

Continual revisions of design assumptions and analysis according to the gaining information about the behaviour of the structure and the rock were carried out.

The observational method is used to control the heading within the influence of the fault zones and therefore the construction sequences of the flexible support system, e.g. to close the gaps of the slots within the TH-steel arch or to apply additional shotcrete to increase the bearing capacity of the support. If the monitoring shows that more favourable ground conditions are present than expected, the observational method may also be used to optimise the rock support, e.g. to reduce the applied support elements.

Due to the use of the observational method the headings of the MFS Faido are intensively monitored with 3D-convergence measuring points, extensometers, measuring anchors with anchors load cells and concrete pressure cells. Borehole rod extensometers measure lengths between four anchor points in a borehole and a reference head at the borehole collar. The relaxation of rock around the tunnel walls can be monitored and therefore the loosening zone can be estimated. The measuring anchor is a combination of an anchor and an extensometer. Its task is to determine the ranges of depths, where the load is taken up due to loosening effects of the rock. It is therefore suitable for the determination of the optimal anchor lengths. Anchor load cells measure the pressure between anchor head and rock mass. Total pressure cells are used to monitor stress changes (radial and tangential) in concrete.

Deformation measurements at the excavation are essential for designing and controlling the tunnel support system. Depending on this information the change from a flexible to a rigid support system can be controlled as well as the application of additional shotcrete at a certain distance behind the tunnel face. The monitoring is necessary in order to understand the soil-structure interaction. The results are regularly compared with the prediction made beforehand. Therefore the rock exploration is regularly investigated by drilling and tunnel seismic prediction (TSP). Both methods are very helpful to locate the difficult/favourable geological sections. The results from the rock investigation and the geotechnical measurements also serve the purpose of calibrating theoretical modelling. This theoretical modelling is an indispensable part of the risk analysis.

Thus the geotechnical computation can be refined as better/more information is available.

5 CONCLUSIONS

With the occurrence of the unexpectedly squeezing ground behaviour the original computations and dimensioning based on rock parameters and the characteristic line curve were carefully analyzed and verified with the help of the Geological Strength index and by back analysis. The results of all considerations and computations show that different methods can lead to completely different results. None of the methods has proved reliable to seize the various influences from the fault zones close-to-reality. Under these conditions, geotechnical computations can give assistance but may not be regarded as sole source for the dimensioning of the support. For the excavation of the further tunnel sections in the MFS Faido, geotechnical computations are complemented with input from the observational method. Important thereby is the current observation of geology on site together with the experiences (observations, measurements, condition support system) from the last excavation steps.

In conclusion, the combination of geotechnical computations and the observational method was successfully used to cross the difficult fault zones in the MFS.

REFERENCES

Flury, St., Zbinden, P. and Roethlisberger, B. (2005). "Challenges encountered during construction of the Faido Multifunctional Station of the Gotthard Base Tunnel and their Solutions", Darmstaedter Geotechnik Kolloqium, 17.03.2005.

Hagedorn, H. (2006). "Hazards during Construction of a Multifunctional Station in a Large fault zone", International Symposium on Underground Excavation and Tunnelling, Bangkok, 02.-04. February 2006.

Dr. Hoek, E. (2000). "Practical Rock Engineering". PDF available at http://www.rocscience.com/hoek/Practical RockEngineering.asp.

Kovári, K. and Lunardi, P. (2000). "On the Observational Method in Tunnelling". PDF available at http://www.igt.ethz.ch/dynDBpages/PublicationDisplay539.htm.

Improving the reliability of GSI estimation: The integrated GSI-RMI system

G. Russo
Geodata SpA, Turin, Italy

ABSTRACT: This paper provides an update on the determination of the Geological Strength Index (GSI, Hoek et al., 1995) by means of a quantitative assessment of the relative input parameters. In particular, an innovative procedure has been found which consists in a rational integration of the GSI with the RMi (Rock Mass index, Palmstrom, 1994). On the base of the conceptual affinity of the GSI with the Joint Parameter (JP), a relationship between the two indexes is derived and exploited in order to obtain a reliable, quantitative assessment of the GSI by means of the basic input parameters for the determination of the RMi (i.e. the elementary block volume and the joint conditions). In this way, the user has the possibility of applying and comparing two truly independent approaches for the determination of the GSI: the traditional qualitative "Hoek's chart", mainly based on the degree of interlocking of rock mass, and the proposed quantitative assessment, mainly based on the fracturing degree of a rock mass. On the basis of such double estimation, a definitive "engineering judgement" can be more rationally expressed. The new approach favourites as well the implementation of the probabilistic approach for managing the inherent uncertainty and variability of rock mass properties. An example of application is presented to illustrate the high potentiality of the method.

1 INTRODUCTION

In current rock engineering practice, it is rather common to use quality indexes for the quantification of the geomechanical parameters, indexes through which the properties of the rock mass are defined, starting from those of the intact rock, taking into consideration the discontinuity network and the relative geotechnical characteristics.

The correct use of such an approach requires in particular:

– a reasonable possibility of assimilation of the rock mass to a "equivalent-continuous" and isotropic geotechnical model;
– reference to "pure" quality indexes that are representative of the geostructural conditions of the in situ rock mass, such as, for example, the "Geological Strength Index" (GSI, Hoek, Kaiser and Bawden, 1995 and following) and the "Joint Parameter" (JP) of the "Rock Mass Index" (RMi, Palmstrom, 1996 and subsequent updates).

As far as the GSI quantification method is concerned, it is worthwhile observing how the Authors initially indicated a derivation from the RMR (Bieniawski, 1973 and following) and Q (Barton, 1974 and following) indexes, after opportune corrections, to take into consideration only the intrinsic characteristics of the rock masses. Later on, however, Hoek progressively abandoned this procedure in favour of a direct determination based only on the use of a diagram ("Hoek's chart", see Fig. 5 later) that summarises the qualitative evaluation of the structural geological characteristics of rock masses and of the relative discontinuity characteristics (Hoek, 1997; Hoek and Brown,1997; Hoek ,1998; Hoek and Marinos, 2000; Hoek, 2005: personal communication).

Furthermore, Marinos et Hoek (2000, 2004) proposed other two diagrams specifically oriented to the determination of the GSI for heterogeneous (such as the flysch) and for very weak (molasse) rock mass, respectively.

The logical aspect of such an evolution is probably related to the objective of having:

– a purely "geostructural" index to reduce the intact property: this is particularly relevant in the case where the source is the RMR, as the uniaxial compressive strength of the intact rock (σ_c) is one of the input parameters;
– a qualitative estimation method that is considered the most suitable for:
  • the classification of the most unfavourable geomechanical contexts (according to Hoek, generally for GSI values <35); incidentally, it can be

observed, at the same time, how for very high GSI values (roughly >75), the use of the index is not recommended for the derivation of the rock mass parameters according to a "equivalent-continuous" model (Hoek , 2005; Diederichs, 2005).

• the evaluation of the "interlockness" degree of the rock blocks;
– a classification method which includes also a wider geological evaluation (Marinos et al., 2004, 2005).

The Hoek's choice has led to a lively discussion at an international level (see for example Stille and Palmstrom (2003), Bieniawski (2004), etc.).

In effect, the basic problem again crops up that has fundamentally favoured the spread of traditional geomechanical classifications, that is the risk of an excessive subjectivity in the estimation by the users, also taking into consideration their different experiences.

Furthermore, the recourse to objective measurements is essential for having a large quantity of data (for example, the borehole core boxes) and to the consequent use of statistical and/or probabilistic analysis. It should be also noted that the evaluation of interlockness degree is often very questionable when examining the core boxes.

On the other side, this last evaluation is probably the most relevant concept introduced by the Author. In fact, it is important to observe that in the Hoek's chart the classification of rock mass structure is not based on the degree of fracturing, but exactly on the interlockness degree of the rock blocks. A practical consequence is that according to the new system the elementary block volume does not necessarily change the assigned GSI rating.

For example, one rock mass should be classified as "Blocky" (Fig. 5) if it is "very well interlocked, consisting of cubical blocks formed by three orthogonal discontinuity sets". This means that in such case, if the discontinuity conditions are not changing, one rock mass formed by cubical blocks of $1 \, cm^3$ will have the same GSI as the one formed by blocks of $1 \, dm^3$, or even of $1 \, m^3$. Consequently, for example, a $10 \, m$ diameter tunnel, subject to a certain stress condition, should exhibit the same excavation behaviour in all these cases.

It is likely to suppose that some practical experiences of excavation behaviour should have convinced the Authors about this concept, which appears to be a very controversial point, because more frequently the common practice seems to support the opposite opinion and, in addition, appears to be in contrast to:

– the most common "pure" indexes for the classification of rock mass quality (RMi, RMrí, Qí, RQD, ...), in which the fracturing degree is one of the main input parameters;

– the results of numerical simulation for example by Distinct Element Method (for example Shen and Barton, 1997; Barla and Barla, 2000);
– the results of laboratory test on samples formed by cubical blocks, which have frequently documented the reduction of the geomechanical properties with the reduction of the block volume; further, it should be added that Barton and Bandis (1982) pointed out that different mechanisms of failure can justify a higher rock mass strength despite the reduction of the unitary block size.

The argument is evidently "tricky" and perhaps some contrasting experience, when not justifiable by different stress conditions or construction procedure, may simply reflect the limit of the "equivalent- continuum" approach, which disregards the intrinsic discontinuity of the rock mass and the actual degree of freedom of the rock blocks with respect to the excavation surfaces.

Taking into consideration the different elements, in favour and against, an approach that adequately integrates both the qualitative and the quantitative assessment appears to be an optimal choice.

2 QUANTITATIVE INPUT FOR GSI ASSESSMENT

Different authors have proposed a quantification of the input parameters for the determination of the GSI, for example, Sonmez and Ulusay (1999) and Cai et al. (2004). All of them maintained the Hoek's chart as a general reference, finding the input criteria to get the same numerical output as the original diagram.

In particular, as schematically outlined in Fig. 1, Cai et al. proposed to use the Unitary Volume of the rock blocks (Vb) and the Joint Condition Factor (JC) as the quantitative input parameters for the determination of the GSI.

As is known, we are dealing with basic parameters for the determination of the RMi index of Palmstrom (1996), even though, in the specific case, the Joint Condition Factor is calculated through the simplified relation of Jc = jW*jS/jA in which jW, jS and jA are the indexes for the quantification of the undulation at a large scale, the roughness and the weathering of the discontinuities, respectively, whose classification points are obtained according to the tables proposed by the Author.

However, the alternative method of keeping completely independent the two possible assessments of the GSI, is here considered preferable, in order to systematically apply and compare:

– the original "qualitative" approach, fundamentally based on the estimation of the degree of interlockness of the rock blocks through the Hoek's chart;

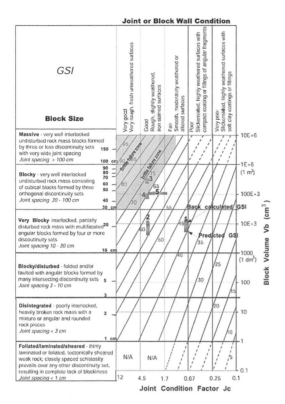

Figure. 1. Hoek's chart (1999) for the determination of the GSI modified by Cai, Kaiser et al., 2004.

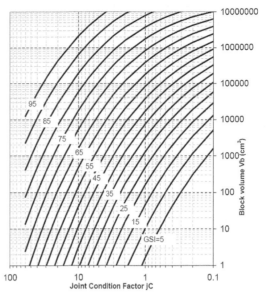

Figure. 2. New proposed diagram for the assessment of GSI by means of the RMi parameters jC and Vb.

– an independent "quantitative" approach, described in the next subsection, centred on the measurement of the fracturing degree of the rock mass.

2.1 The proposed quantitative method: the integrated GSI-RMi system (GRs)

As already mentioned, the existing alternative proposals for the derivation of the GSI are fundamentally centred on the use of the basic parameters of the RMi system, but with adequate modification of the relative weights in order to maintain unchanged the original output (Hoek's chart).

Nevertheless, given the described conceptual background, and in particular the role of the interlockness degree in such a diagram, this exigency appears to be not fundamental and, on the contrary, an alternative and completely independent method is considered more opportune. Such a method is developed taking into consideration the conceptual equivalence between the GSI and the JP parameter (Jointing Parameter) of the RMi system, considering that both are used to scale down the intact rock strength (σ_c) to rock mass strength (σ_{cm}).

According the two systems, we in fact obtain:

1) RMi: $\sigma_{cm} = \sigma c*JP$
2) GSI: $\sigma_{cm} = \sigma c*s^a$ (where s and a are the Hoek & Brown constants)

JP should therefore be numerically equivalent to s^a and given that for undisturbed rock masses (Hoek et al., 2002):

s = exp[(GSI − 100)/9] and a = (1/2) + (1/6)*[exp (−GSI/15) − exp(−20/3)]

The direct correlation between JP and GSI can be obtained, i.e.:

$$JP = [exp((GSI - 100)/9]^{(1/2)+(1/6)*[exp(-GSI/15)-exp(-20/3)]}$$

For the inverse derivation, the perfect correlation ($R^2 = 0.99995$) can be used with a sigmoidal (logistic) function of the type:

GSI = (A1 − A2)/[1 + (JP/X$_o$)^p] + A2

with A1 = −12.19835; A2 = 152.96472; X$_o$ = 0.19081; p = 0.44318.

On the basis of this correlation, a quantitative "robust" estimation of the GSI can be assessed, by defining the parameters concurrent to the evaluation of JP, i. e. the block volume (Vb) and the Joint Condition factor (jC). A graphic representation of the described correlation is presented in Fig. 2.

It should be noted that here the Joint Condition Factor (jC) is of course the original one proposed by Palmstrom, i.e. including the jL factor that expresses the persistence of the discontinuities: jC$_{(Palmstrom)}$ = jR*jL/jA where jR = jW*jS. For

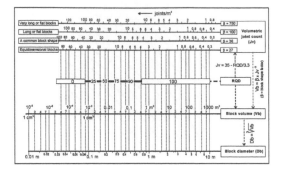

Figure. 3. Different fracturing indexes and their reciprocal correlations (Palmstrom, 2000).

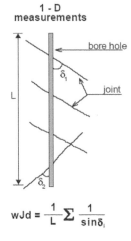

$$wJd = \frac{1}{L}\sum \frac{1}{\sin\delta_i}$$

Figure. 4. Calculation of the wJd from scanline (Palmstrom, 2000).

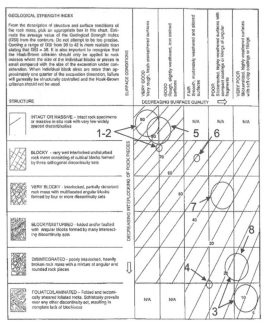

Figure. 5 (a). Some GSI values from different case histories reported in Hoek's papers.

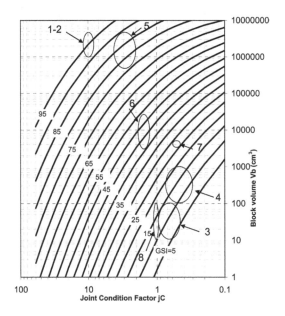

Figure. 5 (b). GSI values obtained for the same case histories as those in Fig. 5(a).

example, the case jL = 1 corresponds to an average joint length of $1 \div 10$ m.

As indicated for example in Fig. 3, Palmstrom developed different methods for the derivation of the Unitary Volume of the Blocks (Vb) on the basis of statistical analyses and illustrated correlations with the different joint indexes of the rock masses (RQD, number of discontinuities per linear, squared or cubic metre (Jv), weighted density of the discontinuities (wJd, Fig. 4), etc.).

The evaluation of the Vb is also improved through the estimation of the shape factor of the rock blocks (β), on the basis of which, for example, the relations $Vb = \beta *Jv^{-3} = \beta *wJd^{-3}$ are proposed, given that, according to the Author, $wJd \approx Jv$.

Furthermore, the Jointing Parameter is calculated by means of the equation $JP = 0.2*jC^{0.5}*Vb^{D}$ in which $D = 0.37*jC^{-0.2}$.

A complete treatment of the RMi method can be found on A. Palmstrom's web site www.rockmass.net.

Just as an example of application, in Figs. 5 (a,b), some case histories reported by Hoek in different

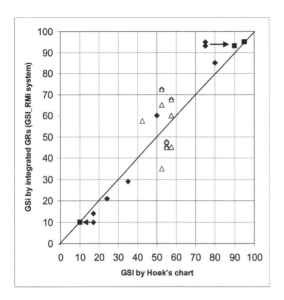

Figure. 6. Correlation between the GSI mean values in Fig. 5 (a,b) and between other practical applications described in the text (triangular symbols). The two arrows in the figure highlight also the effect of some modifications to the original GSI values that appear more consistent with the approximate re-interpretation of the examined data.

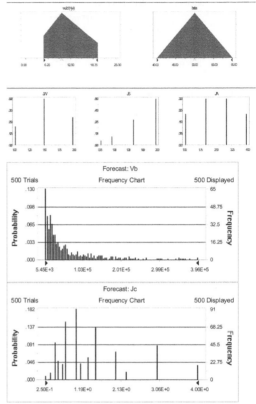

Figure. 7. Example of probabilistic quantitative assessment of GSI. Above: input parameters [wJd(≈ Jv),β ,jW,jS,jA]; below: calculated parameters [(Vb,Jc)]. The derived GSI distributions for both the applied methods are compared in Fig. 9.

papers have been processed for determining the GSI by means of the proposed quantitative approach and compared with the original classification furnished by the Author (Fig. 6).

The link between the considered example and the reference paper is highlighted in the bibliography section by the relative number in square brackets (e.g.: [→3]).

Evidently, this attempt of comparison may be just indicative and in general the evaluation of the discontinuity condition have not been changed from the original in order to focus better on the rock mass structure assessments.

In the diagram of Fig. 6, other direct applications of the two methods to some representative rock outcrops in the Alpine domain have been added for enriching the comparison.

As one can see in such figure, as expectable, a certain difference between the two determinations of the GSI are observed, mainly in the central part of the graph, where probably the influence of the block size rating determines the highest scatter respect the traditional approach.

A comparison between the method proposed by Cai et al. and the new system is shown in the next subsection, by means the application of a probabilistic approach.

2.2 *Probabilistic implementation of the "quantitative" approach*

As already experimented in several practical cases, the application of the described quantitative methods with a probabilistic type of approach is considered to be particularly interesting and of great potential (Russo and Grasso, 2006). This approach allows the variability and/or uncertainty of the available data to be adequately taken into account. In particular, when the latter are statistically significant (in quantitative and qualitative terms), the frequency histograms and/or the density functions that best describe the data distribution are used as input. At the same manner, in cases of great uncertainty and lack of data, the probabilistic approach allows the assumed parametric variability field to be considered on the basis of expert estimates.

Figures 7 and 8(a,b) show an input/output of the probabilistic analysis example conducted applying the

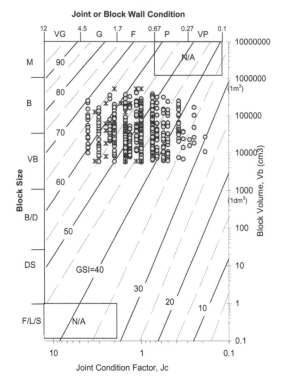

Joint or Block Wall Condition

Figure. 8(a). Results of the probabilistic calculation using the Cai et al. method.

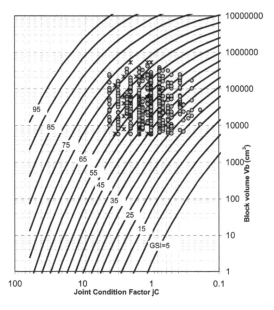

Figure. 8 (b). Results of the probabilistic calculation using the new "GRs" approach.

Table 1. Results of the probabilistic analysis reported in Figs. 8 (a,b).

Fractile	GSI	
	Cai et al.	GRs approach
0.01	33	28
0.25	44	44
0.50	48	50
0.75	54	58
0.99	66	72

MonteCarlo method (500 simulations with Latin-Hypercube sampling) for the probabilistic derivation of Vb and jC, and therefore of the GSI, by the two previously described "quantitative" methods.

In order to favour a comparison between these two methods, a unitary value of the parameter jL is assumed so that Jc $_{(Cai\ et\ al.)}$ = jc $_{(Palmstrom)}$. The analysis examined some surveys performed in calcareous-dolomite rocks and did not consider the fault and/or intense fractured zones, which were studied separately. The results can therefore be considered, in this case, representative of the "ordinary" conditions of the rock mass.

In short, the analysis of the available data led to the quantification of the input parameters with the distributions indicated in Fig. 7 from each of them, at each simulation, a value is sampled and concur to the assessment of a single GSI value. The GSI values obtained from the analysis are explained in the two diagrams shown in Fig. 8 (a,b): each point highlighted by a circle represents a possible result, which is the fruit of the probabilistic combination of the input parameters. For comparison purposes, the graphs also report some deterministic evaluations of the GSI conducted on rock outcrops of the same lithology (cross symbols).

Looking at Table 1 et Fig. 9, it can be seen that, in the case under examination, the use of the two approaches give rather comparable results for the central part of the frequency distributions. The "GRs" approach, however, yields a relatively wider spread in the tails of the distributions, marked by a difference between the two extreme percentiles of 44 points, against the 33 obtained with the Cai method. The simplifying assumption, on one hand, of jL = 1 and therefore Jc = jC should however be recalled and on the other hand, more generally, much more marked differences can be associated to the analysis of more unfavourable geotechnical contexts. It can be seen, for example, how an examination of a hypothetical condition of jC = Jc = 1 e Vb = 10 cm^3 would lead

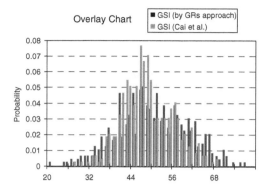

Figure. 9. Overlay chart for the comparison of the results of the probabilistic simulation by the GRs and the Cai et al. approach.

to GSI values equal to about 30 with the Cai method and only about 10 with the GRs.

As commented in section 1, it is interesting to observe that the use of the Hoek's chart might lead to very high GSI values also in such highly fractured conditions of the rock mass, if, for example, the "Blocky" structure would be recognised.

3 CONCLUSIVE REMARKS

A new method for the GSI estimation has been proposed, mainly based on the quantitative assessment of the same parameters concurring to the calculation of the Jointing Parameter (JP) used for the determination of the RMi.

The approach is not intended to substitute the "qualitative" approach centred on the use of the Hoek's chart, but more properly to integrate it by a completely independent system.

In such a way, the final engineering judgement can be assessed on the basis of both the traditional method, essentially based on the degree of interlocking of rock masses, and the new system, mainly based on the measured state of fracturing.

Furthermore, the new approach is not covering special cases of complex and/or weak rock, in which the cited specific charts proposed by Hoek and Marinos appear to be more adequate, if the basic conditions of applicability of the GSI are satisfied.

Finally, as further important step, it is important underline that the proposed approach favourites as well the concurrent calculation of the RMi and consequently the possibility of application of the empirical method for tunnel design developed by Arild Palmstrom.

Given the complementarities, the proposed integrated system appears to be very promising.

REFERENCES

Barla G. and Barla M. (2000): *Continuo e discontinuo nella modellazione numerica dello scavo di gallerie* Gallerie e Grandi Opere Sotterranee n. 61.

Barton N., Lien R. and Lunde J. (1974): *Engineering classification of rock masses for the design of tunnel support.* Rock Mechanics, vol.6, n. 4.

Barton N. and Bandis S. (1982): *Effects of block size on the shear behavior of jointed rock.* Proc.23th Symp. Rock Mech. Berkeley.

Bieniawski Z.T. (1973): *Engineering classification of Jointed Rock masses.* Trans. South African Inst. of Civil Engineers. Vol.15, No.12, pp. 335–344.

Cai M., Kaiser P.K., Uno H., Tasaka Y. and Minami M. (2004): *Estimation of Rock Mass Deformation Modulus and Strength of Jointed Hard Rock Masses using the GSI system.* International Journal of Rock Mechanics and Mining Sciences n.41, pp. 3–19.

Diederichs M. (2005): *General Report for Geodata: Methodology for spalling Failure and Rockburst Hazards.*

Hoek E. (2005): Personal communication.

Hoek E. and Brown E.T. (1980). *Underground Excavations in Rock.* The Institution of Mining and Metallurgy, London.

Hoek E. and Brown E.T. (1997). *Practical estimates of rock mass strength.* Submitted for publication to the Int. J. Rock Mech. Min. Sc.& Geomech. Abstr. [→ *examples n. 1,2,4*]

Hoek E. and Diederichs M. (2005): *Estimation of rock mass modulus.* International Journal of Rock Mechanics and Mining Science (in press).

Hoek E. and Marinos P., (2000): *Predicting Squeeze.* Tunnels and Tunneling International, November, pp.45–51.

Hoek E., Carranza-Torres C. and Corkum B. (2002): *Hoek-Brown failure criterion – 2002 Edition.* Proc. North American Rock Mechanics Society. Toronto, July 2002.

Hoek E., Kaiser P.K. and Bawden W.F. (1995): *Support of Underground Excavations in Hard Rock.* Balkema, Rotterdam, 215pp.

Hoek E., Marinos P. and Benissi M. (1998): *Applicability of the Geological Strength Index (GSI) classification for very weak and sheared rock masses. The case of the Athens Schist Formation.* Bull.Eng.Geol.Env. 57(2) 151–160. [→ *example n. 3*].

Marinos P., Hoek E., Marinos V. and, (2004): *Variability of the Engineering properties of rock masses quantified by the Geological Strength Index. The case of Ophiolites with special emphasis on tunneling.* Proceed. of Rengers Symposium. [→ *examples n. 5,6,7,8*].

Marinos P., Marinos V. and Hoek E., (2004): *Geological Strength Index, GSI: Applications, recommendations, limitations and alteration fields commensurately with the rock type.* Bull. of the Geolo.Society of Greece vol XXXVI- Proc. of the 10th Intern.Congress, Thessaloniki.

Palmstrom A. (1996): *Characterizing rock masses by the Rmi for use in pratical rock engineering* Tunn. and Und. Space Tech. vol.11.

Palmstrom A. (2000): *Recent developments in rock support estimates by the RMi.* Journal of Rock mechanics and tunneling technology, Vol.6, pp. 1–9.

Russo G. and Grasso P. (2006): *Un aggiornamento sul tema della classificazione geomeccanica e della previsione del comportamento allo scavo* Gallerie e Grandi Opere Sotterranee n. 80

Shen B. and Barton N. (1997): *The disturbed zone around tunnels in jointed rock masses.* Int.J.Rock.Mech.Min. Sci.Vol. 34

Sonmez H. and Ulusay R., (1999): *Modifications to the Geological Strength Index (GSI) and their applicability to stability of slopes.* Intern.Journal of Rock Mchanics and Mining Sciences n.36, pp. 743–760.

Underground Works under Special Conditions – Romana, Perucho, Olalla (eds)
© 2007 Taylor & Francis Group, London, ISBN 978-0-415-45028-7

Effect of lower seam old workings on longwall powered supports in upper seam

V.R. Sastry & R. Nair
National Institute of Technology Karnataka, Surathkal, Mangalore, India

M.S. Venkat Ramaiah
Singareni Collieries Company Limited, Kothagudem, Andhra Pradesh, India

ABSTRACT: Load bearing capacity of powered roof supports is a key issue in the design of longwall panels. The existence of multiple seams in a given project may involve over mining or undermining operations across different seams. The paper envisages variation in stress distribution and load on powered supports in the upper seam longwall panel due to the presence of already mined out bottom seam consisting goaf, pillars and openings through detailed field investigations in an underground coal mine followed by finite element modeling study. Studies revealed that for every 40 m of face retreat, the load from the roof gradually gets transferred from the panel to the goaf. Stress distribution over chock shields was considerably affected when the face in upper seam reached 15–25 m zone above the openings in lower seam. The presence of barrier in the bottom seam increases load on the chock shields in longwall panel of upper seam.

1 INTRODUCTION

Depleting mineral reserves together with steep rise in oil prices necessitate the exploitation of deep seated coal deposits. Occurrence of multiple seams is a common feature in Indian coal mining projects. The lower quality deposits occurring in upper seam lead to the mining of better quality lower seams by bord and pillar technology in some of the Indian coal mines. The upgradation of technology and the changed economic scenario have made the mine operators to extract coal in upper seams existing above already worked out lower seams. It is, therefore, essential to study the influence of old workings like barriers and goaf on the stress distribution in the longwall panels of the upper seam.

In design of mine layouts, stability of the workings due to extraction of a particular seam as well as the effects from already extracted seams should be accounted (Chugh & Pytel 1998). The roof falls during extraction and subsequent compaction of goaf tend to disturb the insitu stress in floor or roof, i.e. the parting between the two seams (Brady & Brown 1992). The stability of the parting between two seams is also important for successful longwall extraction in multiple seams. To understand the inter-seam effects during mining process, numerical modeling can be utilized by simulating field conditions and analyzing the results in the form of stress distribution.

The stress envelope formed during extraction process may extend to the underlying/ overlying seam causing large variation of stress in the affected strata. Thus, the stress variation affects the loading of pillars, powered supports and longwall face. In case of designing upper seam longwall workings operating in multi-seam environment, the location and magnitude of stress transferred from lower seams has to be determined and then the effect of stress transferred from lower seams on stability of workings in upper seams can be predicted (Lou 1997). For creating a numerical model of overmining condition, compaction of underlying goaf with respect to bulking factor has to be considered (Trueman 1990). A comprehensive investigation programme was taken up to study the effect of underlying seam workings on the longwall panel operating in the upper seam by detailed field monitoring in the longwall panel and also by finite element modeling.

The upper seam longwall panel was extracted with following conditions present in the lower seam workings at different locations along panel length:

1. Goaf of lower seam bord and pillar workings
2. Virgin area, where coal remained unextracted
3. Openings

Due to the importance of powered roof supports in longwall machinery, the load behavior over the supports was considered in detail during field investigations.

2 INVESTIGATIONS

Studies were carried out in an underground coal mine in Southern India. Details of the mine are given below:

2.1 About the mine

The mine where the investigations were carried out is in Kothagudem Area of Singareni Collieries Company Limited, Andhra Pradesh, India, situated at the southern part of Godavari Valley coal fields. The mine is located approximately between latitude 17°29' 13" N to 17°30' 35" N and longitude 80°39' 28" E to 80°41' 0" E. Two workable seams existed in the mine (Table 1). Thickness of the top seam is varying from 6 to 10 m. The seam is dipping at 1ine due N68°E. Presently, top 3 m section of the upper seam is extracted by mechanized longwall retreating method. The longwall workings extend from a depth of 59 m on the rise side and 234 m on the dip side. The roof at working face is supported by 4×760t chock shields of IFS type with extension bar. Bottom seam contains high quality coal and has been extensively developed and extracted in two sections, i.e. along the floor and roof with a parting of 4 m in the middle, by bord and pillar method.

2.2 About the panel

Panel-A was worked at a depth of 215 m with dimensions of 420 m × 150 m. Three boreholes were drilled at 40 m, 80 m and 180 m along the length of the panel up to the depth of the working section, which indicated no major changes in the stratigraphic and physico-mechanical properties of the overlying strata. The overlying roof beds in the panel were medium to coarse grained sandstones with intercalation of shale and sandstones. The RQD of the roof beds was varying from 45% to 94%, whereas compressive strength was varying from 8.5 to 10.5 MPa. The overlying beds are massive in nature, but their caving indices range from easily cavable to moderately cavable. Below the longwall panel of top seam, old workings of bord and pillar extraction are present in bottom seam with a partition of 45 m (Table 1). Between 215 m to 345 m face retreat, variation in bottom seam working condition existed both along length of longwall panel and also along face of the panel.

3 FIELD MONITORING

Extensive field instrumentation programme was taken up to assess the loading pattern on longwall panel, for the entire length of 440 m, during extraction using various instruments (Fig. 1). Leg pressure surveys were conducted by fixing pressure gauges on each leg of all the 102 chock shields and continuous pressure recorders at strategic chock shield legs (Figs. 2 & 3). Objective of this was to understand the loading of front and rear legs of chock shields as the face retreated. Data was generated for entire length of the panel. A Multi-point Borehole Extensometer with two anchor points was installed at a distance of 30 m from the barrier by drilling a borehole from the surface along

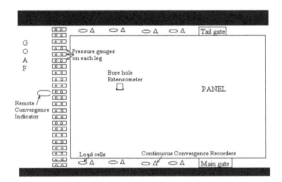

Figure 1. Positions of monitoring instruments in panel.

Table 1. Thickness and parting of the seams.

Seam	Description		Thickness (m)
Top seam	Roof	Sandstone with shale and carbonaceous clay intercalation	2.14 to 10.2
	Floor	Sandstone	–
	Parting	Sandstone, shale, shaley coal with thin coal bands	42.0 to 45
Bottom seam	Roof	Sandstone	3.57 to 9.45
	Floor	Sandstone	–
	Parting	Sandstone	5.0 to 6.0

Figure 2. Continuous pressure recorders at chock shield no. 50 and 75.

132

the central line of the panel, the aim of which was to observe the strata movement in the goaf. The above recordings were used to validate finite element model.

Leg pressure surveys were conducted by pressure gauges fitted to each leg of all chock shields and by continuous pressure recorders at chock shields legs to understand the load transfer onto the powered supports. The continuous pressure recorder was attached to chock shield no. 25, 50 and 75. The pressure recording from the gauges were monitored thrice in a day, whereas continuous pressure recorders gave data on 24 hour basis. The plot obtained from continuous pressure recorders was helpful in recording the variation of pressure and monitoring of leg bleeding due to excessive load.

The pressure recordings on the chock shields for the entire length of the panel were converted to mean load densities (Fig. 4). The peaks shown in Figure 4 indicate the occurrence of roof falls (weightings) in the panel during extraction. The increase of mean

load densities beyond 72 m of face progress may be attributed to the excessive load on the chock shields due to the cantilever effect of the hanging roof. The pressure distribution was not uniform during main fall along the length of the face (Fig. 5). There was clear distinction of pressure values between upper half of the face and lower half of the face from chock shield no. 43. The pressure towards the tailgate side from middle of the panel was less than the pressures on the main gate side. The observed behavior can be attributed to influence of underlying workings. The tail gate side was lying over goaf of bottom seam, whereas the main gate was lying above the virgin section of bottom seam. Hence, it appears that the distressed condition of rock mass, due to the already worked out sections below the tail gate, is the reason for the pressure reduction over chock shield legs in the tail gate area. The legs of chock shields lying above the virgin/barrier pillar of the lower seam experienced high stresses as could be seen in Figure 5. Further, it was found that there was equal loading on front and rear legs during main fall.

Continuous pressure recorders indicated a lesser variation of leg pressures in chock shield no. 25 compared to chock shield nos. 50 and 75 (Fig. 6). The legs of chock shield no. 25 attained yield pressures of 38 MPa only twice as compared to 17 and 15 times for chock shield nos. 50 and 75 respectively. This was due to the fact that the chock shield no. 25 was traveling above virgin bottom seam for entire face retreat, whereas chock shield nos. 50 and 75 were traversing above different bottom seam conditions.

Multi-point Borehole Extensometer (MPBEx) was installed from surface at 30 m from the installation chamber. To monitor the strata movement, two anchor points were fixed at 40 m and 50 m into the roof of the working section. Recordings indicated that anchor at 40 m above the roof got detached after a face retreat of 52.2 m and the anchor point at 50 m above the roof got detached after a face retreat of 57 m. Since these anchors were fixed above the main roof, it can be

Figure 3.   Pressure gauges on each leg of chock shields.

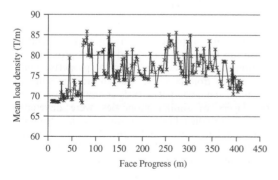

Figure 4.   Mean load density for the entire length of panel.

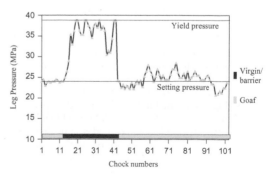

Figure 5.   Pressure distribution on chock shields during main fall.

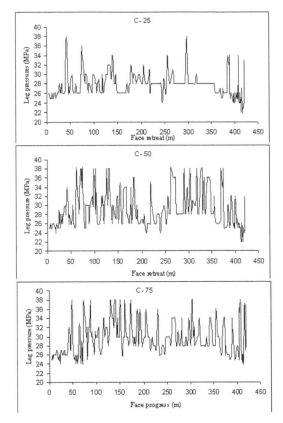

Figure 6. Pressure survey using continuous pressure recorders.

concluded that the main weighting has taken place before a face retreat of 50 m. The same observation was made in modeling study and it validated the veracity of the finite element model.

4 FINITE ELEMENT MODELING

Models were prepared for different positions of long-wall face over the bottom seam mined-out workings, using finite element software – NISA. The entire rock mass was modeled using 3-D Hexahedron elements. The extent of the model was limited to 240 m in X direction (along length of the panel) and 120 m in Y direction (along depth) which includes both the seams (Fig. 7). In the model, 60 m of overburden was considered above the top seam. A vertical loading of 3.3 MPa was applied to upper nodes (calculated by considering the average density of overburden rock mass as 22 kN/m^3), in order to simulate the remaining overburden of 150 m from surface. The boundary elements were modeled as roller supports. In the vertical boundaries on both sides, a relief of 10 mm was

Table 2. Multi point bore hole extensometer recordings.

Average face progress (m)	Reading (mm)	
	Anchor-1 at 172 m depth (40 m above roof)	Anchor-2 at 162 m depth (50 m above roof
22.70	0.14	0.42
23.60	0.14	0.42
24.60	0.14	0.42
26.10	0.14	0.42
28.30	0.28	0.82
30.00	0.48	0.92
31.00	0.72	1.46
33.00	3.40	1.48
35.10	14.75	1.48
37.40	18.54	1.58
40.60	26.47	1.82
42.20	55.50	1.92
44.70	71.11	1.96
46.60	120.00	2.12
49.50	136.00	2.28
50.70	236.00	2.61
52.20	Anchor lost	180.6
54.60		240.00
56.9		Anchor lost

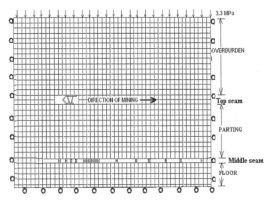

Figure 7. Meshing of the model.

provided for displacement in horizontal direction, since the boundary considered was in proximity to workings. The physico-mechanical property inputs were taken from borehole data. In order to simplify the input data, composite beds comprising of different layers of similar properties were developed according to the laboratory results obtained from the borehole data. Table 3 and 4 give the details of properties of the composite beds, considered in the model. According to the mine plan, variations in bottom seam condition exist between 215 m to 345 m face retreat, both along the longwall panel length and also

Table 3. Input parameters for roof of the top seam.

Strata	Youngs modulus (MPa)	Poissons ratio	Cohesion (MPa)	Angle of internal friction (q)	Mass density (kN/m3)
Coal Seam	2×10^3	0.3	0.012	14	14
Bed 1	3.4×10^3	0.32	1.9	39	20.50
Bed 2	2.6×10^3	0.35	1.9	39	20.40
Bed 3	2.6×10^3	0.35	1.9	39	20.40
Bed 4	3.5×10^3	0.4	1.9	39	20.60
Bed 5	2.6×10^3	0.35	1.9	39	20.40
Bed 6	3.6×10^3	0.32	1.8	12	20.50
Bed 7	3.4×10^3	0.32	1.9	39	20.50
Bed 8	3.4×10^3	0.32	1.9	39	20.50
Bed 9	3.4×10^3	0.32	1.9	39	20.50
Bed 10	3.4×10^3	0.32	1.2	14	20.50

Table 4. Input parameters for floor of the top seam (Parting).

Strata	Youngs modulus (MPa)	Poissons ratio	Cohesion (MPa)	Angle of internal friction (θ)	Mass density (kN/m^3)
Coal floor	6.25×10^3	0.3	0.012	14	22.50
Bed 1	4.3×10^3	0.32	1.9	32	21.60
Bed 2	1.97×10^3	0.35	1.9	32	22.40
Bed 3	1.42×10^3	0.35	0.19	32	21.00
Bed 4	3.6×10^3	0.32	0.19	32	21.80
Bed 5	3.6×10^3	0.32	0.012	16	21.80
Bed 6	3.6×10^3	0.32	0.012	16	21.80
Goaf	1.3×10^3	0.1	–	–	15.00
Rib	2.31×10^3	0.3	0.012	14	14.00
Barrier	6.25×10^3	0.3	0.012	14	22.50

along the face of the panel. The bottom seam conditions like barriers, openings and goaved out areas were incorporated into the model. The goaf compaction in the bottom seam working was taken as 80%, since the bottom seam was worked out 40 years ago.

Chock shield pressure of 0.7 MPa was modeled, considering 70% of the yield pressure of the legs as setting pressure. An unsupported span of 1 m was provided in the front of canopy tip simulating the immediate roof condition encountered after a slice of coal is extracted by shearer.

Since the rock mass behaves nonlinearly with the loading conditions, an elasto-plastic model was considered with Drucker Prager criteria to define the yield behavior. A strain hardening parameter was also incorporated in the model in order to simulate the behavior of stress- strain curve after failure. The rock mass was first allowed to stabilize from the insitu stress thus incorporating virgin condition in the subsequent runs. Model was then run for varied conditions in the lower seam.

5 MODELING RESULTS

The model was prepared in the Pre-processor – DISPLAY IV and further analysis was done with the core of NISA II, both being packaged into a single general software of Numerically Integrated elements for System Analysis (NISA). Output from the analysis was obtained in terms of horizontal stress, vertical stress and shear stress. The predicted results from numerical modeling were considered at the following points (Fig. 8):

1. Behind the chock shields (point 1)
2. Over the chock shields (middle) (point 2)

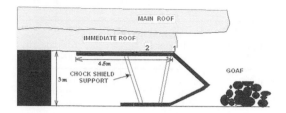

Figure 8. Monitoring stations in the model.

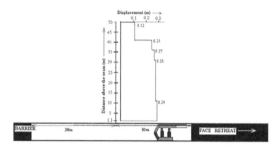

Figure 9. Strata movement from the model analysis during face retreat.

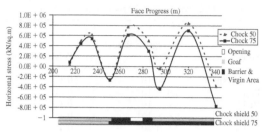

Figure 10. Variation of horizontal stress behind the chock shield during face progress.

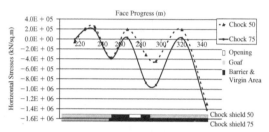

Figure 11. Variation of horizontal stress middle of the chock shield during face progress.

The model was run at different predetermined intervals, which was deduced on the basis of the position of longwall face over the bottom seam condition. The longitudinal section was considered on 2 chock shields i.e. chock shields no. 50 and 75 along the panel length.

5.1 Analysis of bed separation

Bed separation above the seam being extracted was analyzed from the model at the position where the multipoint borehole extensometer was installed. The strata movement at mainfall was recorded as shown in Figure 9. The analysis showed that after a retreat of 50 m of longwall face, i.e. after crossing MPBEx point by 20 m, the strata at 42 m above in the roof of the working section recorded a displacement of 0.25 m, indicating breakage of roof leading to mainfall. The data obtained from modeling study corresponds closely to the borehole extensometer recordings wherein the anchor at 42 m above the roof of longwall face showed a strata movement of 0.23 m. This study has indicated the validity of the model.

5.2 Analysis of the horizontal stress

The modeling study results of development of horizontal stress pattern around the working face are given below:

1. The horizontal stress distribution behind the chock shields did not show any change between face retreat from 215 m to 265 m (Fig. 10). There is a

continuous increase of stress magnitude in each cycle, which indicates the stress build up in the roof strata. This phenomenon also indicates the interval of periodic falls (Fig. 11). The stress over chock shield no. 50 shows marginal decrease in stress value as compared to chock shield no. 75 between 265 m to 280 m face retreat, indicating the effect of opening in the bottom seam. At 295 m retreat of the face in upper seam, an increase of horizontal stress in chock shield no. 50 by about 3 MPa was observed, which is due to solid barrier existing below chock shield no. 75, whereas chock shield no. 75 was lying above the goaf.

2. Similarly, stress distribution on the middle of the chock shields indicates that the variation in stress is similar in case of all the chock shields (Fig. 11). The presence of opening in bottom seam caused a sudden reduction in horizontal stress at middle of the chock shield no. 50. The stress magnitude on chock shield no. 50 reduced by 60% due to the opening in the bottom seam between 270 m to 280 m of face retreat. This decrease of horizontal stress was also observed for further face progress from 280 m to 300 m.

As the longwall panel progressed from goaf to virgin area, there was an increase in stress observed both behind and middle of the chock shields. The effect of opening in the bottom seam on horizontal stress distribution over the chock shield was observed from

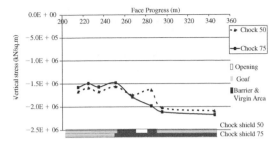

Figure 12. Variation of vertical stress on behind the chock shields during face progress.

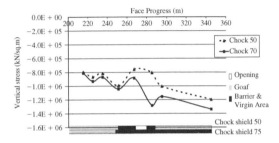

Figure 13. Variation of vertical stress on middle of the chock shield during face progress.

about 25 m before the working face was actually above the opening.

5.3 *Analysis of the vertical stress*

Similarly, observations of vertical stresses on the chock shields were made with model studies and major findings are as follows:

1. Insignificant variation in vertical stress was found behind both the chock shields between 220 m to 255 m retreat of longwall face, since the chocks were lying above goaf in the bottom seam (Fig. 12). Marginal decrease in stress behind the chock shield no. 50 was observed as face moved over the opening of lower seam from one end to other. As the longwall face in the top seam retreated above the underlying bottom seam goaf to barrier, increase of vertical stress of 0.3 MPa behind the chock shield was observed. Vertical stress was observed to be less on chock shield no. 50 than on chock shield no. 75 during face progress beyond 290 m, since goaf existed below chock shield no. 50 whereas chock shield no. 75 was lying over solid barrier.

2. Over the chock shields (in middle portion), an increase and decrease of stresses was observed for every 40 m interval (Fig. 13). Increase of stress was observed when the top seam longwall face moved

above the underlying goaf to barrier. In case of chock shield no. 50, decrease in load was observed from 270 m to 288 m. This was due to the presence of a large opening in the bottom seam which resulted in increase in the vertical stress on the chock shields. Decrease in stress on middle of the chock shield no. 50 was observed from 248 m face retreat, which was lying above the barrier in the bottom seam till a maximum reduction at 275 m face retreat, at which point in the face was lying above the centre of the opening in bottom seam. Subsequently, the vertical stress increased as the face retreated beyond 275 m. The effect of the bottom seam opening was observed on the stress over the chock shield no. 50 till 295 m of longwall face retreat in upper seam. It can be inferred that the presence of opening in bottom seam affects stress distribution over the chock shields when the top seam longwall face moves in the zone of 20 m before the opening to 15 m beyond the opening.

Model studies indicated increase in vertical stress on the chock shields, as the longwall face in upper seam traversed above the bottom seam goaf to the virgin zone. There was a reduction in horizontal and vertical stresses on the chock shields in top seam due to the opening present in the bottom seam. The stress distribution on the chock shields in upper seam longwall panel was affected when it retreated along the horizontal zone of 20 m before the bottom seam opening and 15 m after the opening. There was no considerable change in vertical stress when the longwall face was above the goaf in the bottom seam as compared to the condition when the longwall face was lying above the barriers in bottom seam. This may be indicative of considerable compaction of goaf that took place over the years. Results are also indicate that the parting of 45 m is sufficient for the safe working of top seam longwall panel above the old goaved out zones in lower seams. The model study revealed a regular increase and decrease of stress over the chock shields, corresponding to the phenomenon of stress build up and stress relief at regular interval, indicating periodic falls in the panel.

6 CONCLUSIONS

The study revealed a marginal increase of load on the chock shields when the longwall face traverses from goaf to virgin condition in bottom seam or vice versa. Presence of opening in the bottom seam was found to affect the stress distribution over the chock shields in the top seam longwall panel. For the particular geomining conditions encountered in the mine, it can be concluded that the parting of 45 m between two seams is sufficient for the safe working of the panel with 7×460 t powered supports in upper seam.

REFERENCES

Brady, B.H.G. & Brown, E.T. 1992. Rock mechanics for underground mining. Chapman and Hall, London, 2nd edition, ISBN 0 412 47550 2.

Chugh, Y.P. & Pytel W.M. 1998. Design of partial coal mine layouts for weak floor strata conditions. Proceedings of the workshop on coal pillar mechanics and design, 33rd U.S Symposium of Rock Mechanics, June 7.

Luo, J. 1997. Gate road design in overlying multi-seam mines, M.S. Thesis, Virginia Polytechnic Institute and State University, Blacksburg, Virginia, USA.

Trueman, R. 1990. A finite element analysis for the establishment of stress development in coal mine caved waste. In. Journal of Mining Science and Technology, 10: 247–252.

Underground Works under Special Conditions – Romana, Perucho, Olalla (eds)
© 2007 Taylor & Francis Group, London, ISBN 978-0-415-45028-7

Modelling and validation of ground behaviour in a very old tunnel rehabilitation

V. Navarro Torres & C. Dinis da Gama
Technical University of Lisbon

S. Longo
Geotechnical Center of IST

ABSTRACT: The human development history has always registered tunnel constructions for different uses, a trend that is increasing in the recent years, because they induce great advantages in the economic, social and environmental aspects.

Parallel to the intensification of tunnel constructions is the growing need to develop various types of interventions such as maintenance, repairing and rehabilitation, for improving their sustainability along time.

In Lisbon, due to prolonged time spans of service provided by the Rossio railway tunnel, which was built between 1887 and 1889, it was recently decided to carry out significant rehabilitation interventions..

The paper describes the modelling and validation of "in situ" instrumentation results regarding the behaviour of a low quality ground, located in a 17.50 m long segment of that tunnel, during the rehabilitation works for new primary support application.

1   LOCATION OF ROSSIO TUNNEL AND THE STUDY AREA

The Rossio railway tunnel connects the Lisbon center to the outbound station of Campolide, situated 2.6 km away. The tunnel has been subjected to important

rehabilitation works which include a segment located between Pk 0 + 350 and Pk 0 + 367.50 m that has caused several stability problems both within and outside the tunnel (Fig. 1).

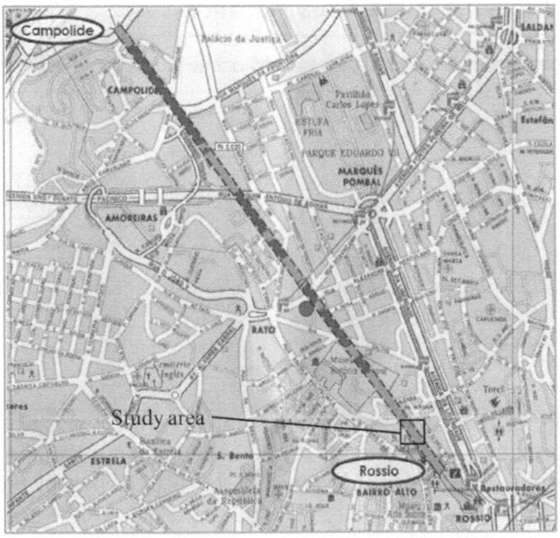

The direction of Rossio tunnel is N45°W and it has 2.613 km of length with a 8.0 m × 6.0 m cross section.

The 17.50 m tunnel segment that was selected for this presentation is situated in a highly sensitivity area because it is located below a health center called "Centro das Taipas" (Fig. 2).

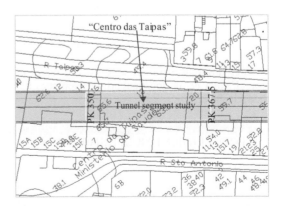

Figure 1.   Layout of Rossio tunnel in the center of Lisbon and location of its segment from Pk 0 + 350 to Pk0 + 367.5 m.

Figure 2.   Detailed location of tunnel segment under study.

Table 1. Local works during initial tunnel construction.

Local works	Pk	Distances of local works (m)	Location
Rossio	0 + 194	0	Axes
Shaft I	0 + 470	276	Axes
Shaft II	0 + 687	217	Axes
Shaft III*	1 + 178	491	Axes
Shaft IV*	1 + 588	410	10 m of axes
Shaft V	2 + 283	695	12 m of axes
Shaft VI	2 + 768	485	Axes 69.92 m
Campolide	2 + 807	38	Axes

* For ventilation.

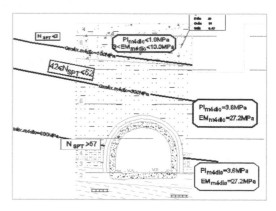

Figure 3. Geologic formations at Pk 0 + 350 to 0 + 367.5 segment 1: Marly clay, 2: Clay with silt, 3: Sand with silt, 4: Clay with silt, 5: Silts with fine sands and clay with silt, 6: Fill (REFER, 2006).

2 HISTORY OF TUNNEL CONSTRUCTION

The construction of Rossio tunnel started in July 25, 1887 and was concluded in May 24, 1888, taking a total of 11 months (GRID, 2005).

The tunnel has a general slope of 1% towards the Campolide portal and it was built under the so-called Belgium construction method, which includes the following steps: excavation of an advanced gallery at the central and the upper sections; widening of the upper laterals; construction of roof liner with brick masonry; expansion of the section's central part; widening of the lower laterals and finally construction of supports at the lowest part of walls with bricks and limestone rock in areas of soil and weathered rocks.

Besides the two portals at Rossio and Campolide six shafts were opened for both ventilation and construction of separate tunnel segments at an advance average rate of 12 meters per day (Table 1).

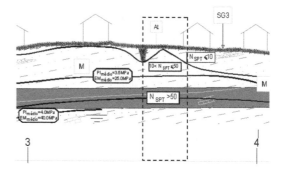

Figure 4. Longitudinal section of Pk 0 + 350 to 0 + 367.5 area (REFER, 2006).

3 GEOTECHNICAL CHARACTERISTICS

The geological composition at the tunnel segment Pk 0 + 350 to 0 + 367.5 includes several Miocene sedimentary layers. The stratigraphic sequence is composed of 6 sub-horizontal formation types (Fig. 3).

The geotechnical properties of the Miocene strata above tunnel roof are very poor in terms of strength (Fig. 4).

Three different materials were used to represent ground strata at the location, whose properties are shown in Table 2.

Table 2. Geotechnical strata specifications.

Formation	Fill	Miocene **	Clay***
γ (kN/m^3)	20	20	21
E (MPa)	60	40	60
ν	0.35	0.35	0.35
c (MPa)	0.30	0.10	0.10
ϕ (°)	35	30	28

* Indicated with number 1 in the following figures.
* Indicated with number 2.
* Indicated with number 3.

4 FINITE ELEMENT MODEL SIMULATIONS

4.1 Model characteristics

The tunnel segment rehabilitation was simulated using a finite element model to check the numerical values of measured displacement obtained inside the tunnel and at the surface. A computer model was prepared on the basis of a 15 m total extension in the longitudinal direction. To calculate the displacement field in greater detail a more realistic model should be generated, with at least three times length, with the tunnel segment between Pk 0 + 350 and Pk 0 + 367.5 at its centre.

Table 3. Input data for FE model geometry.

Model dimensions		
Width (m)	Height (m)	Length (m)
60	45	17
Old tunnel internal dimensions		
Width (m)		8.6
Height (m)		6.6
Brick walls and crown width		0.8
New tunnel internal dimensions (m)		
Width (m)		12.2
Height (m)		9.3
Micropoles length (m)		4
Rockbolts length (m)		8
Overburden* (m)		12.8

* From the crown of the old tunnel to surface.

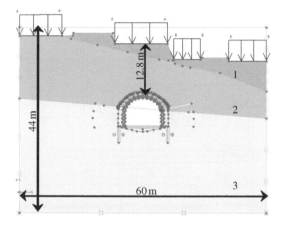

Figure 5. Input model cross-section. Three different geologic formations are indicated, according to Table 2. The figure shows also surface applied loads, as well as old and new tunnel support systems.

However, with the 15 m length model all project characteristics were dully considered (geometry of the old tunnel of the enlarged new tunnel, geology, water table level, surface buildings and tunnel construction phases). Micropiles, rock bolts, steel arches and steel fiber reinforced shotcrete were modeled according to the specifications of the materials really used.

Geometry of the model is presented in Table 3.

The presence of surface buildings were simulated with a pressure of 40 kN/m² distributed loads applied in three different positions for all cross-sections (Fig. 5).

Water table level reached the base of the old tunnel as shown in Figure 6.

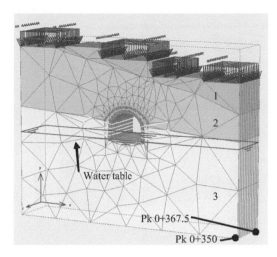

Figure 6. 3D view of the model, with rockbolts installed (staged construction). Loads and water level are visible. The mesh was refined only around and above the tunnel.

4.2 *The simulation of construction phases*

The finite element software allowed to closely simulate all construction phases in which the real tunnel enlargement and rehabilitation was executed. In this model 24 phases were analyzed.

The first 3 phases were necessary to obtain the surface topography, to apply the loads and to substitute the old tunnel lining for rock. In phase 4 rock bolts were introduced for the whole length of the model as well as micropiles, for a length of 12 m. Phases 5 to 11 were processed to open one meter of tunnel enlargement at each phase, from the front of the model (Pk 0 + 350) to Pk 0 + 357. In phase 5 no support system was applied, while in phases 6 to 11 steel fiber reinforced shotcrete, steel arches and micropiles were installed, one meter behind the tunnel enlargement heading.

Tunnel floor was maintained intact from phases 1 to 22. At phase 12, one more arch of micropiles was set, from Pk 0 + 358 to the last section of the model. A 5 m length micropile superimposition was respected as in the real tunnel rehabilitation. In phases 13 to 22 the tunnel enlargement continued, and phase 22 contains the introduction of support system for the arch and the walls in the last meter that was opened. Phases 23 to 27 were introduced to remove the old tunnel floor and build the floor of the new tunnel in reinforced concrete.

In phase 23, 4 m of tunnel floor were removed and no support system was installed. Phase 24 considered an additional advance of 4 m at the excavation floor, and the first 4 m were supported with steel fiber reinforced shotcrete. The last phase was generated to simulate the applied support to the last 4 m of floor in the enlarged tunnel.

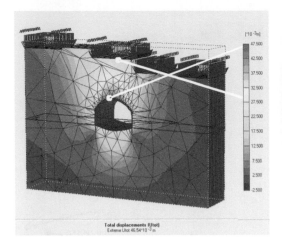

Figure 7. Vertical displacement field in the first section of the whole model.

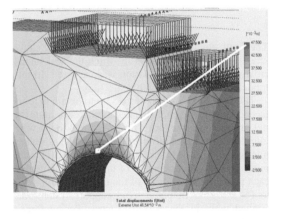

Figure 8. Expanded view of vertical displacements at tunnel roof and at the surface.

4.3 *Obtained results*

More realistic results are visible in the first transversal sections, due to its non optimal extension in the longitudinal direction. Fig. 7 shows vertical displacements in the first cross-section of the whole model where the vertical displacement of the enlarged tunnel roof reached 4.6 cm inwards. Subsidence reached 2.6 cm, vertically above the tunnel segment from a minimum of 3 mm.

Fig. 8 shows the vertical displacement distribution at the upper part of the model. Displacements in the tunnel floor reached 3.5 cm inwards, approximately at middle distance between the tunnel axis and the walls (Fig. 9).

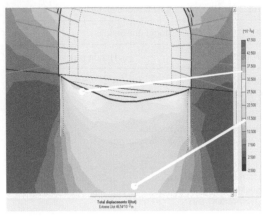

Figure 9. Zoom view of tunnel floor with vertical displacements.

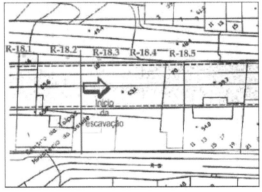

Figure 10. Location of points for measuring displacements.

5 IN SITU MEASURED DISPLACEMENTS

5.1 *Measurement of displacements at "Centro das Taipas" building*

The systematic recording of settlements was conducted at the "Centro das Taipas" building located above the tunnel segment by means of monitoring movements at five points (Fig. 10).

The obtained values of settlements in that building reached between 18 and 22 mm in the stage of the primary support process application and later on values of 20 to 26 mm were obtained (Fig. 11), thus corresponding to largely observed cracking and abundant damages.

5.2 *Convergence measurements in the tunnel*

The results of convergence measurements inside the tunnel at the section under study reached maximum

R-18.3 (PK 0+355,6)

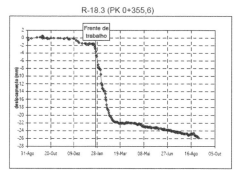

R-18.4 (PK 0+364,4)

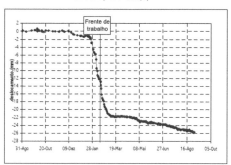

R-18.5 (PK 0+371,9)

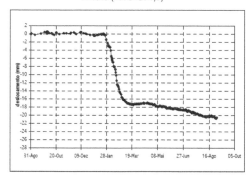

Figure 11.   Results of settlement measurements at "Centro das Taipas" building (Navarro V. 2006).

values of 30 mm in section 2–3 and of 12 to 16 mm in the 1–2 and 1–3 sections (Fig. 12).

Comparing modeling results with field measurements leads to the conclusion that the displacements inside the tunnel are relatively different than numerical predictions, unlike those in the building, which were quite similar. The explanation of these deviations arises from difficulties and problems of the construction process, which were often detected and documented.

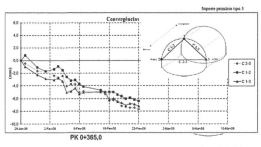

PK 0+365,0

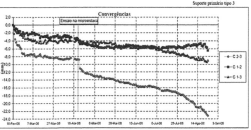

Figure 12.   Tunnel convergence measurement results as a function of time (Navarro, 2006).

6   CONCLUSIONS

The study has shown that it is extremely important to monitor tunnel constructive processes through both "in situ" measurements and numerical modeling, in order to interpret anomalies of that activity as well as understanding their causes, and also to control upper ground behavior during excavation for preventing damages to the surrounding structures.

In the case study that was described, this interesting procedure allowed the confirmation of inadequate constructive techniques practised by the contractor company, which motivated high displacements at the surrounding ground and caused expensive damages to adjacent buildings.

REFERENCES

GRID, 2005. Design, reinforcement and rehabilitation of the Rossio Tunnel. Compilation and synthesis of available information (PCRRTR-CSID) – Volume I: Data summary, Lisbon (in Portuguese).
REFER, 2006. Geological and Geotechnical Study, 2nd Version, Lisbon (in Portuguese).
NAVARRO TORRES, V. 2006. Study of the geotechnical behavior at the Centro das Taipas area during the rehabilitation process of the Rossio tunnel, Lisbon (in Portuguese). Geotechnical Center of IST and DHV/FBO Consultants, S.A.

Underground Works under Special Conditions – Romana, Perucho, Olalla (eds)
© 2007 Taylor & Francis Group, London, ISBN 978-0-415-45028-7

Instability assessment of a deep ramp influence on the surrounding exploitation openings in a copper mine

C. Dinis da Gama & V. Navarro Torres
Geotechnical Center of IST, Technical University of Lisbon

ABSTRACT: The well known influence of new openings over surrounding cavities in underground mines is particularly important in deep operations.

This paper addresses the instability assessment of a ramp called CRAM03 in the Neves Corvo copper mine located from depths 760 m to 827 m and constructed in a volcanic rock mass with massive mineral, which influenced the surrounding underground openings of the ore extracted by Drift and Fill, and Bench and Fill exploitation methods.

The adopted methodology is described and consisted firstly of the rock mass geotechnical characterization in interaction with the underground openings, followed by the identification of the instability level, based on standards obtained through in situ test results, and finally by developing prevention techniques to control their instability risk.

1 INTRODUCTION

Rock mass strength was assessed in the present study by the Mohr–Coulomb and Hoek- Brown failure criteria in order to analyze stability conditions of two areas of the Neves Corvo mine: ramp CRAM03 with a 5 m × 5 m section and Bench and Fill stopes of North Neves area. The rock types were mainly formed by filite and quartzite (PQ) and the volcano-sedimentary complex (CVS).

The uniaxial compressive strength of the PQ rock groups varies from 70 to 100 MPa and of the CVS group from 160 to 200 MPa, with an average specific mass of 3200 kg/m3 (National Laboratory of Civil Engineering, LNEC, 1987).

The mean RMR index (Rock Mass Ratio) for these rock types is 70, and the Poisson's ratio value is around 0.20 (varying from 0.15 to 0.30). The Young's modulus varies from 31.6 to 40 MPa. (J. Lobato, 2001).

2 ROCK MASS FAILURE CRITERIA

2.1 *Mohr–Coulomb failure criterion*

In general terms, this criterion considers failure to be attained by applying to a rock volume an axial stress σ_1 under a confining stress of σ_3. It is given by equation (1) involving the uniaxial compressive strength of the

$$\sigma_1' = k\sigma_3' + \sigma_{cm} \tag{1}$$

rock mass σ_{cm} and a k factor:

The k value is the declivity obtained when representing σ_1 in function of σ_3 and it depends on the friction angle of the rock mass φ, by:

$$k = tg^2(45^o + \phi/2) = \frac{1 + sen\phi}{1 - sen\phi}.$$

The uniaxial compressive strength of the rock mass σcm is given by equation (2), in function of the cohesion c and the internal friction angle:

$$\sigma_{cm} = \frac{2c\cos\phi}{1 - sen\phi} \tag{2}$$

2.2 *Hoek–Brown failure criterion*

This failure criteria is based on the principal stresses σ1 e σ3 (Hoek, E.; Brown, E.T., 1982), in function of

Table 1. Typical values of the m and s failure criteria parameters (Hoek and Brown, 1982).

Rock type	(1)		(2)		(3)		(4)		(5)	
RMR	m	s	m	s	m	s	m	s	m	s
100	7.0	1.0	10.0	1.0	15.0	1.0	17.0	1.0	25.0	1.0
85	3.5	0.1	5.0	0.1	7.5	0.1	8.5	0.1	12.5	0.1
65	0.7	0.004	1.0	0.004	1.5	0.004	1.7	0.004	2.5	0.004
44	0.14	0.0001	0.2	0.0001	0.3	0.0001	0.34	0.0001	0.5	0.0001
23	0.04	0.00001	0.05	0.00001	0.08	0.00001	0.09	0.00001	0.13	0.00001
3	0.007	0	0.01	0	0.015	0	0.017	0	0.025	0

(1): Carbonated and crystallized rocks (dolomites, limy, marble, etc.)
(2): Argillaceous rocks (clay, limonite, shale, etc.)
(3): Well cemented sands rocks (sandstone, quartzite, etc.)
(4): Igneous rocks of fine grain (andesite, diabase, riolites, etc.)
(5): Igneous rocks with thick grains (Gneiss, granite, quartz-diorite, etc.)

the uniaxial compressive strength of the intact rock and two coefficients m and s depending on rock properties (Table 1).

$$\sigma_1 = \sigma_3 + \sigma_{cm}\left(m\frac{\sigma_3}{\sigma_{cm}} + s\right)^{0.5} \quad (3)$$

When there is no lateral confinement ($\sigma_3 = 0$) and s = 1 (laboratory tests), it results $\sigma_1 = \sigma_{cm}$.

The tensile strength σt is obtained by considering $\sigma_1 = 0$ and $\sigma_3 = \sigma_t$, providing the following equation:

$$\sigma_t = 0.5\sigma_{cm}\left(m - (m^2 + 4s)^{0.5}\right) \quad (4)$$

Also for rock mass classification, a very useful index is Q the Quality Index, proposed by Barton, N., et al., (1982), and expressed by equation (5) where, RQD is the Rock Quality Designation (Deere, D. U., 1964), J_n is the number of joint sets , J_r is the joint roughness value, Ja is the alteration degree of joint walls, J_w is the water presence value and SRF is the stress reduction factor, as a function of the "in situ" state of stress:

$$Q = \left(\frac{RDQ}{J_n}\right)\left(\frac{J_r}{J_a}\right)\left(\frac{J_w}{SRF}\right) \quad (5)$$

The correlation between Q index and the tunnel representative dimension (D_e) is expressed by Barton's equation (6) and illustrated in Figure 1, where the critical curve (D_{ec}), and separates supported and unsupported conditions.

$$D_e = \frac{L}{ESR} \quad (6)$$

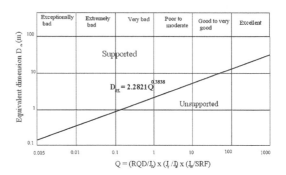

Figure 1. Geomechanical classification according to Q index and support requirements (based in Barton, N., et al., 1982).

Where L is the width, diameter or height of the underground opening (m) and ESR is Excavation Support Ratio, which in the present paper was considered between 1.6 and 2.0.

3 GEOTECHNICAL INSTABILITY IN CRAM03 RAMP AND NORTH STOPE

3.1 Characterization of study area

The Neves Corvo mine is located in the so called "Iberian Pyrite Belt", 230 km Southeast of Lisbon. In geotechnical terms the CRAM03 ramp (Fig. 2) is located at 660 m depth and the surrounding rock mass is good to very good, with zones of medium to bad quality (Table 2).

The Neves North stopes (Fig. 3) are located between 760 m and 823 m depth; and the corresponding mining method is by Drift and Fill, with 5 m × 5 m of

Table 2. Geotechnical characteristics of convergence stations in CRAMS03 ramp (Navarro Torres, V., 2003).

Rock	Station	RQD	Q	RMR	ECU	RMS
CVSsup	3	43–85	4–7	55–63	60–100	34–45
CVSinf	5	30	3–10	52	70	31
PQ	11	40–90	12–23	66–75	50–170	62–96

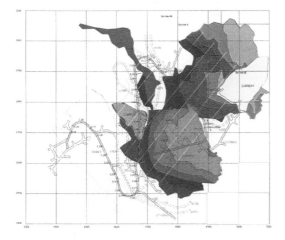

Figure 2. Localization of CRAM03 ramp with respect to adjacent stopes (J. Lobato 2001).

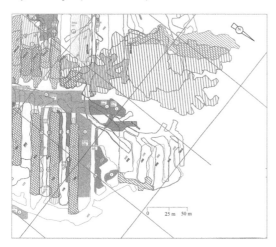

Figure 3. Neves North stope area (Navarro Torres, Vidal et al., 2003).

opening size, and by Bench and Fill method with height varying between 16 to 30 m.

The rock masses in the Neves North stope are formed by two types: volcanic tuffs with fissured minerals and sulphides (Mf) and massive sulphildes (Sm). Studies of Golder in 1994 (Bench and Fill Mining of Neves Orebody) indicate that the UCS of the intact rock is from 160 to 210 MPa and according to LNEC varies from 130 to 220 MPa. Other properties are: Q index is around 9, Young's modulus of 20 GPa, Poisson's ratio 0.2 and the average value of RMR is 65.1.

3.2 Geotechnical instability assessment in CRAM03 ramp

In situ measurements carried through from 1992 to 2000 allowed the application of geotechnical assessment criteria for determining the instability level in the CRAM03 ramp (Table 3) using $\Delta\sigma$/RMS values that range from −5 to 24 (Table 4).

In CRAM03 ramp there were 19 total stations, and it does not exist instability in 8 stations (42%). Thus, it has low level of geotechnical instability in 8 stations (or 42%), moderate in one station (5%) and high level in two stations (or 9%) of the total.

3.3 Geotechnical instability assessment in Neves North stope

With "in situ" measurements carried out from 1992 to 2000 it were elaborated a geotechnical assessment criterion for application in Neves North stope based in safety factors Fs (Table 5).

Table 3. Geotechnical instability assessment criteria in CRAM03 ramp.

Instability level (IL)	Rock mass type	
	CVS $\Delta\sigma$ /RMS	PQ $\Delta\sigma$ /RMS
Low ▽	$5 \leqslant \Delta\sigma$ / RMS < 10	$6 \leqslant \Delta\sigma$ / RMS < 12
Moderate ⊗	$10 \leqslant \Delta\sigma$ / RMS < 23	$12 \leqslant \Delta\sigma$ / RMS < 24
Height ◆	$\Delta\sigma$ /RMS $\geqslant$ 23	$\Delta\sigma$ /RMS $\geqslant$ 24

Table 4. $\Delta\sigma$/RMS and IL values obtained for CRAM03 ramp in 2001 and 2003.

Station	550-CI	C2	C4	C6	C7	C9	C11	C12	C15	C17
$\Delta\sigma$/RMS	−5	0	−5	9	4	9	9	5	10	11
IL				▽		▽	▽	▽	⊗	▽
Station	C20	C24	C27	C28	C31	C32	C34	C36	C37	
$\Delta\sigma$/RMS	6	1	1	3	6	1	10	28	30	
NI	▽				▽		▽	◆	◆	

147

In the Neves North stope the prediction of geotechnical instability for 2001 was low (Fs = 2) and moderate for 2003 (Fs = 1.6) (Table 6).

In the Neves North stopes, for the referential year of 2001, the prediction was for low geotechnical instability levels in S52, S10 and S81 stopes, and in 2003 moderate levels for the S33, S29, S60D, S42 and S1 stopes.

3.4 Corrective actions for the geotechnical instability

The support system selection for the underground openings with moderate to high geotechnical instability levels was based on the Q index (determined after the RMR index) and also on the cavities equivalent dimensions De (Table 7).

For the support system selection, the GDA (Geomechanical Design Analysis) software was utilized, with an ESR factor value of 1.8 (see Fig. 4).

That software examines the stability of underground openings of various geometries as tunnels, drifts, galleries, ramps and stopes in underground mines (User Manual for GDA, 2000).

This program includes analysis of rock mass classification with RMR, GSI, Q and Q' criteria, determination of rock mass properties for numerical modelling, analysis and design of cable bolting systems; stress analysis in longitudinal openings and stopes of diverse shapes; stress sensitivity analysis due to a sequence of advancing openings considering shape,

size and orientation; wedges stability analysis in 3D and support system selection.

The simulation results for the CRAM03 ramp (Fig. 4) indicate that it is not necessary to apply supports, but for the same areas where geotechnical instability risks exist bolting with 2.4 m length and 20 mm of diameter is recommended, and for the Neves North stopes systematic bolting with 3.5 m length and 20 mm of diameter and spacing of 2.5 m is advisable.

Table 7. Rock mass index for support system selection.

Index	CRAM03 ramp		
Local	C15	C36	C37
RMR	68.9	72.1	73.5
Q	20.0	17.5	20.0
D_e	2.80	2.80	2.80
Index	Neves North stopes		
Local	S33 + S29 + S60D + S42 + S1		
RMR	65.1		
Q	9.0		
D_e	4.60		

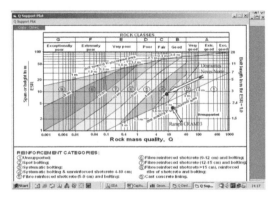

Figure 4. Determination of corrective actions with GDA software.

Table 5. Geotechnical instability assessment criteria of in Neves North stope.

	Rock mass type	
Instability level (IL)	Sm	Mf
Low ▽	3 ≥ Fs > 2	3 ≥ Fs ≥ 1.5
Moderate ⊗	2 ≥ Fs > 1.5	1.5 ≥ Fs > 1.0
Height ◆	Fs = 1.5	Fs = 1.0

Table 6. IL obtained for Neves North stope in massive sulphur Sm for 2001 and 2003.

Year	Mining stopes	Pillars in the Bench and Fill mining method			
		Pillar resistance (MPa)	Pillar stress (Mpa)	Fs	NI
2001	S52 + S10 + S81	70	35	2.0	▽
2003	S33 + S29 + S60D + S42 + S1	70	45	1.6	⊗

4 CONCLUSIONS

For assessing underground openings geotechnical instability is essential to involve in the analyses all inherent geotechnical parameters affecting failure and stability of the surrounding rock masses.

A referential standard for determining the allowable level of geotechnical instability may be based on "in situ" measurements results carried along several years.

The geotechnical instability predictions for underground openings evaluated on the basis of the referential standard and the future ground control techniques allows the application of selected adequate supports, which constitutes the procedure for implementing a pro-active and preventive program, instead of corrective actions.

The models presented in this paper on the assessment of geotechnical instability under special conditions are thus very useful in the sustainable management of underground mining workings.

ACKNOWLEDGEMENTS

The authors would like to thank Mr António Correia de Sá and Mr. José Lobato for their effective contributions and support for the study.

REFERENCES

Barton, N., et al. (1980). "Estimation of support requirements for underground excavations", 16th Symposium on Design methods in Rock Mechanics. Univ. Minnesota.

Deere, D.U., (1964). "Technical Description of Rock Cores for Engineering Purposes", Rock Mechanics and Engineering Geology. Volume 1.

Hoek, E., and Brown, E.T. (1982). "Underground Excavations in Rock", Institution of Mining and Metallurgy, London.

Lobato, J. (2000). "Evaluation of Mining Induced Stress on Support Requirements at Neves Corvo". Master of Science dissertation at the University of Exeter.

Navarro Torres, V. (2003). Environmental Underground Engineering and Applied in Portuguese and Peruvian Mines. PhD Thesis, IST Lisbon.

User manual of GDA-Geomechanical Design Analysis for Underground Openings, 2000. Dias Engineering Inc. Sudbury, Ontario, Canada.

Underground Works under Special Conditions – Romana, Perucho, Olalla (eds)
© 2007 Taylor & Francis Group, London, ISBN 978-0-415-45028-7

Calculation of the pre-stressed anchorage by 3D infinite element

Y.F. Wang, Y.H. Wang & H.Y. Xie

School of Civil Engineering and Mechanics, Huazhong University of Science & Technology, Wuhan, China

ABSTRACT: Based on one dimension infinite element theory, the coordinate translation and the shape function of 3D point radiate 8-node and 4-node infinite elements are derived. They are respectively coupled with 20-node and 8-node finite elements to compute the compression distortion of the pre-stressed anchorage segment. The results indicate that when the pre-stressed force act on the anchorage head and segment, the stresses and the displacements in the rock around the anchorage head and segment concentrate on the zone center with the anchor axis and they decrease with exponential forms. Therefore, the stresses and the displacement spindles are formed. The calculating results of the infinite element are close to the theoretical results. That indicates the method of this paper is right. This paper introduces a new method to study the anchor mechanism of stresses anchors. The obtained results have important role in the research of the anchor mechanism and engineering application.

1 INTRODUCTION

In the anchor system, the reinforcement is acted by two ways. One way is to compact the rock mass through the anchor head. The other way is to deliver the pre-stress to the plasm by the adhesive effect of the cable and the plasm (which is the pull-type anchor) or by the bearing plate (which is compact-type anchor), and then the pre-stresses are delivered to the rock mass around the anchorage segment by the plasm. The head and the segment of the pre-stress anchor are key parts by which the anchors force act on the rock mass and anchor it. Fig. 1 shows a modal configuration of an anchorage head, and Fig. 2 shows a modal configuration of a pull-type anchorage segment.

At present, various numerical methods have been widely applied to the theoretical study of the anchorage design and calculation analysis. The finite element method, such as ANSYS and FINAL, and the finite difference method, such as the fameous FLAC have been used more. Because the case of the anchors bearing pre-stresses belongs to a semi-infinite field, the infinite element method which can preferably simulate the infinite semi-field is applied to in this paper. The infinite element method is developed from the finite element method. It is accurate to resolve the finite field issue by the finite element method. When the finite element method resolves the infinite field issue, the model with enough big finite field and manual handling boundary is adopted to simulate the far region effect. But the precision of this method is not high (Zhang, 2005; Xiong, 2005). For the simulation field is bigger, the calculation workload is more. At the same time, the infinite boundary can not be right reflected. The finite element method is extended to the analysis of the infinite field by the infinite element method. The infinite element method can reflect the boundary condition that the displacement of the

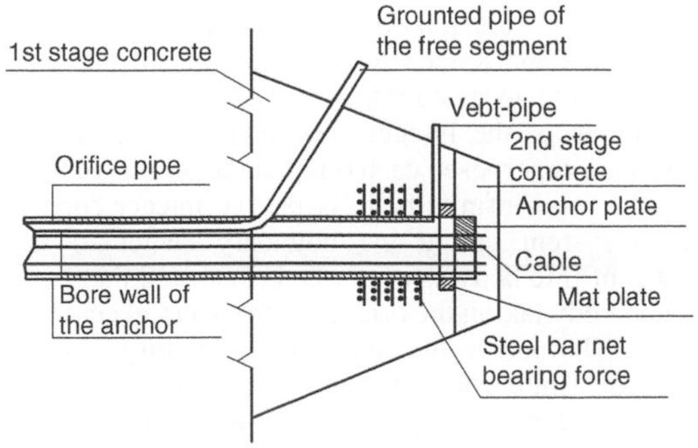

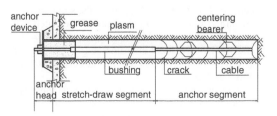

Figure 1. Modal of the head of the great tonnage anchor.

Figure 2. Modal of the centralized pull-type anchor.

infinite distance is zero, and it can right reflect the stress of the anchor head and segment.

The concept of the infinite element was first present by R. Ungless (1973). Then it had been improved and developed by Bettess (1980), Beer (1981), Zienkiewicz (1983). In China, Ge (1986), Zhao (1986 & 1988) had devoted to the development of the infinite element method. At present, the infinite element method has been developed from one dimension to three dimension, from single way mapping to multi-way mapping. Thereinto, the three dimension multi-way mapping infinite element is a new element discussed and developed by scholars. The three dimension models of 8-node and 4-node dot radiation infinite element are respectively coupled with 20-node and 8-node three dimension finite element, and the stress of the anchor head and segment are analyzed in this paper.

1.1 *Introduction and development of the radiation functions of the infinite element*

The concept of the coordinate radiation was introduced in the infinite element method by Bettess (1980). He pointed out that the infinite element and the finite element can use the same coordinates functions, but the displacement shape functions of the two element were different. The variable character of the infinite distance can be reflected by leading a decaying function $f(\xi)$ in the infinite element.

$$M_i = f(\xi)N_i(\xi) \tag{1}$$

where M_i denotes the shape function of the infinite element, and N_i denotes the shape function of the corresponding finite element.

Ziekiewicz (1983) put forward a new infinite element. In one dimension case, the coordinates transform was given as

$$x = N_0(\xi)x_0 + N_2(\xi)x_2 \tag{2}$$

where,

$$N_0(\xi) = \frac{-\xi}{1-\xi}, \quad N_2(\xi) = \frac{\xi}{1-\xi} + 1 \tag{3}$$

When $\xi = 1$, both $N_0(\xi)$, $N_2(\xi)$ have the singularity.

Assumed $x_1 = (x_0 + x_2)/2$, it can be validated that $\xi \to +1$, $x \to \infty$; $\xi = 0$, $x + x_2$; $\xi = -1$, $x = x_1$. Namely, when the local coordinates $\zeta = -1, 0, +1$ respectively correspond to the integer coordinates x_1, x_2 and ∞ in Fig. 3.

In fact the previous radiation takes x_0 as the mapping pole and it has two nodes x_1, x_2. Because x_1, x_2 are not mutually independent, assumed $x_0 = 0$, then nothing but x_1 the radiation from the local coordinates

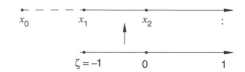

Figure 3. 2-node mapping for 1D infinite element.

Figure 4. 1-node mapping for 1D infinite element.

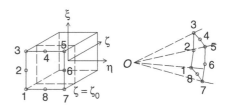

Figure 5. 3D point-radiate 8-node infinite element.

to the integer coordinates can be come true, namely, it is the single node radiation. The coordinates mapping function of the single node radiation can be written as

$$x = N(\xi)x_1 = \frac{2}{1-\xi}x_1 \tag{4}$$

the local coordinates $\zeta = -1, 0, +1$ respectively correspond to the integer coordinates x_1, $2x_1$ and ∞ (shown in Fig. 4).

1.2 *Mapping function of the 8-node dot radiation infinite element*

Figure 5 shows the three dimension radiation extending from the previous one dimension radiation. We can regard the 8-node three dimension infinite element of the local coordinates as eight single node radiations of the integer coordinates. All radiation poles are superposed each other at the origin. To handle the semi-infinite body issue, the integer coordinates system taking the boundary center of the semi-infinite body as origin is formed, and the semi-round body taking the origin as the spherical center is taken. The part in the spherical face is dispersed by 20-node finite elements, and the other outside the spherical face is dispersed by 8-node infinite elements. The radiation pole of the infinite element is the spherical center, and the nodes are on the spherical face. The local coordinates extend to infinity distance. The coordinates radiation are given as

$$x = \sum_{i=1}^{8} M_i x_i , \quad y = \sum_{i=1}^{8} M_i y_i , \quad z = \sum_{i=1}^{8} M_i z_i \qquad (5)$$

It can be assumed that the ζ direction is infinite in three dimension. The coordinates of a random point in the $\zeta = \zeta_0$ cover is only formed by the coordinates interpolation of eight peak on the cover, the shape function corresponding to η, ξ is invariability on the cover. Because the cover joined with the finite elements must satisfy the continue condition, the interpolating functions of the eight nodes on the $\zeta = \zeta_0$ cover can be given as

$$M_i = \frac{1}{4}(1 + \xi \xi_i)(1 + \eta \eta_i)(\xi \xi_i + \eta \eta_i - 1) \quad (i = 1,3,5,7)$$

$$M_i = \frac{1}{2}(1 + \eta \eta_i)(1 - \xi^2) \qquad (i = 2,6) \quad (6)$$

$$M_i = \frac{1}{2}(1 + \xi \xi_i)(1 - \eta^2) \qquad (i = 4,8)$$

Synthesizing the formula (4) and (6), the interpolating functions of the three dimension 8-node infinite element are derived as following:

$$M_i = \frac{(1 + \xi \xi_i)(1 + \eta \eta_i)(\xi \xi_i + \eta \eta_i - 1)}{2(1 - \zeta)} \quad (i = 1,3,5,7)$$

$$M_i = \frac{(1 + \eta \eta_i)(1 - \xi^2)}{1 - \zeta} \qquad (i = 2,6) \quad (7)$$

$$M_i = \frac{(1 + \xi \xi_i)(1 - \eta^2)}{1 - \zeta} \qquad (i = 4,8)$$

Assumed that at the infinite distance the displacement field attenuates from the spherical boundary, the attenuation center is the origin of the integer coordinates. Because all nodes of the dot radiation infinite elements locate at a spherical face, it can be approximately represented as

$$\frac{r_0}{r} = \frac{1 - \zeta}{2} \qquad (8)$$

The radius of the spherical boundary is r_0. The distance from the calculation point outside the spherical boundary to the origin is r. Then the displacement attenuation functions can be expressed as

$$f(\zeta) = (\frac{1 - \zeta}{2})^\alpha \quad \alpha \geq 1 \qquad (9)$$

The displacement transform formulas are

$$u = \sum_{i=1}^{8} N_i u_i , \quad v = \sum_{i=1}^{8} N_i v_i , \quad w = \sum_{i=1}^{8} N_i w_i \qquad (10)$$

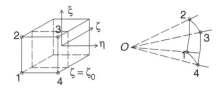

Figure 6.   3D point-radiate 4-node infinite element.

Where N_i are shape function as follows:

$$N_i = \frac{(1 + \xi \xi_i)(1 + \eta \eta_i)(\xi \xi_i + \eta \eta_i - 1)}{4}\left(\frac{1 - \zeta}{2}\right)^\alpha (i = 1,3,5,7)$$

$$N_i = \frac{(1 + \eta \eta_i)(1 - \xi^2)}{2}\left(\frac{1 - \zeta}{2}\right)^\alpha \qquad (i = 2,6) \qquad (11)$$

$$N_i = \frac{(1 + \xi \xi_i)(1 - \eta^2)}{2}\left(\frac{1 - \zeta}{2}\right)^\alpha \qquad (i = 4,8)$$

1.3   Mapping function of the 4-node dot radiation infinite element

In accordance with the upper, the mapping functions of the three dimension 4-node dot radiation infinite element can be obtained (shown as Fig. 6).
The coordinates radiation functions of the 4-node infinite element are given as

$$x = \sum_{i=1}^{4} M_i x_i , \quad y = \sum_{i=1}^{4} M_i y_i , \quad z = \sum_{i=1}^{4} M_i z_i \qquad (12)$$

It can be assumed that the ζ direction of the three dimension is infinite. The interpolating functions of the four points on the $\zeta = \zeta_0$ cover are

$$M_i = \frac{1}{4}(1 + \xi \xi_i)(1 + \eta \eta_i) \quad (i = 1,2,3,4) \qquad (13)$$

Synthesizing the formula (4) and (13), the interpolating functions of the three dimension 4-node dot radiation infinite element are obtained as

$$M_i = \frac{(1 + \xi \xi_i)(1 + \eta \eta_i)}{2(1 - \zeta)} \quad (i = 1,2,3,4) \qquad (14)$$

The interpolating functions of the 8-node and the 4-node infinite elements both satisfy the following rules.

(i)   When $\zeta = -1$, $\Sigma M_i = 1$, and it is corresponding to the interpolating functions of the finite elements on the cover.
(ii) When $\zeta \rightarrow +1$, $M_i \rightarrow \infty$.
(iii) When $i \neq j$, $M_i(\zeta_i, \eta_i, \xi_i) = 0$; when $i = j$, $M_i(\zeta_i, \eta_i, \xi_i) = 1$.

153

The displacement transform formulas are

$$u = \sum_{i=1}^{4} N_i u_i, \quad v = \sum_{i=1}^{4} N_i v_i, \quad w = \sum_{i=1}^{4} N_i w_i \quad (15)$$

where N_i are the displacement functions are given as

$$N_i = \frac{(1+\xi\xi_i)(1+\eta\eta_i)}{4}\left(\frac{1-\varsigma}{2}\right)^{\alpha} \quad (i=1,2,3,4) \quad (16)$$

The displacement functions of the 8-node and the 4-node infinite elements satisfy the following rules.

(i) When $\zeta = -1$, $\Sigma N_i = 1$, and it is corresponding to the interpolating functions of the finite elements on the cover.
(ii) When $\zeta \to +1$, $N_i \to 0$, namely, the displacement all go to zero. It satisfies the condition that the displacement of the infinite distance is zero.
(iii) When $i \neq j$, $N_i(\zeta_i, \eta_i, \xi_i) = 0$; when $i = j$, $N_i(\zeta_i, \eta_i, \xi_i) = 1$.

Base on the coordinates radiation functions and the displacement transform functions, followed the element stiffness matrixes of the point radiation infinite elements, the usual process of the finite elements can be derived. The author programs to analyze the stresses of pre-stress anchors.

2 STRESS CALCULATION OF THE ROCK AROUND THE ANCHORS

It can be assumed that assume that the pre-stress force of an anchor is 200 kg, and the Poissonís ratio and Youngís modulus of the rock are respectively $\mu = 0.23$ and $E = 2 \times 10^4$ MPa. The pre-stress of the anchor head is evenly distributing in the region of 1 m × 1 m. The pre-stresses are straight acted on the node of the anchorage segment top.

2.1 Stress calculation of the rock around the anchor head

Based on the symmetry the quarter model is calculated. The spherical boundary is a spherical face whose radius is 4 m. The element mesh is shown in Fig. 7. The finite element method coupling with the infinite element is adopted. The body in the boundary is dispersed by twenty three dimension 20-node solid finite elements, and the boundary is dispersed by twelve three dimension 8-node infinite elements. We order α in the formula (14) of the displacement attenuation function equals to 0.

The results are shown in Fig. 8 and Fig. 9.

Fig. 8 and Fig. 9 show the results computed by the infinite element method. Fig. 10 shows the results

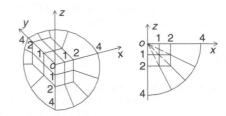

Figure 7.   Mesh of the mode of the anchorage head.

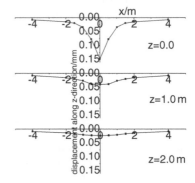

Figure 8.   Displacements of the anchorage head by the infinite method.

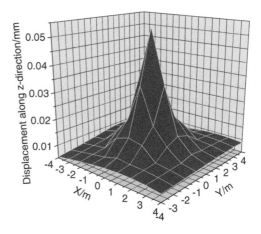

Figure 9.   Three-dimension graphs of displacements around the anchorage head (z = 1.0 m).

computed by the infinite element method and the theoretical method derived by the author (it is demonstrated in other paper). The following characters of the displacements of the rock around the anchor head can be obtained.

154

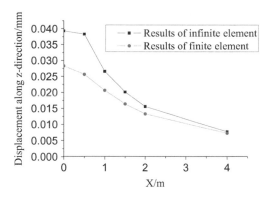

Figure 10.　Constraction between the results of the infinite element and the theory.

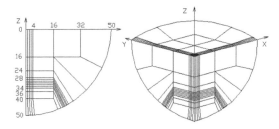

Figure 11.　Mesh of the mode of the anchorage segment.

(i)　The displacements of the rock around the anchor head mostly focus on a zone center with the origin. The radius and the depth of the zone are respectively 2 m and 2 m.

(ii)　The displacements decrease with exponential forms. Thus, a displacement taper around the anchor head is formed. That indicates the stresses shape of the rock around the anchor head is taper.

(iii)　The results computed by the infinite element method is approximate to the results computed by the theoretic method derived by the author. Only near the anchor, the results computed by the infinite element method is bigger than the results computed by the theoretic method. The reason is the pre-stress used in the infinite element method is translated into a fixate on a node, but the force used in the theoretic method is an integral of the uniform load. So the results curves of the theoretic method are more evenly.

2.2　Tress calculation of the rock around the anchor segment

Based on the symmetry, the quarter model is calculated. The spherical boundary is a spherical face whose radius is 50 m. The element mesh is shown in Fig. 11. The method coupling the finite element with the infinite element is adopted. The body in the boundary is dispersed by 434 three dimension 8-node solid finite elements, and the boundary is dispersed by 148 three dimension 4-node infinite elements. We order α in the formula (14) of the displacement attenuation function equals to 0.

The results are shown in Fig. 12 and Fig. 13. In the two figures, the values of z begin from the force action spot.

Fig. 12 and Fig. 13 show the results computed by the infinite element method. Fig. 14 and Fig. 15 show

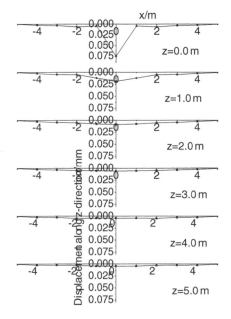

Figure 12.　Displacements of the anchorage segment by the infinite method.

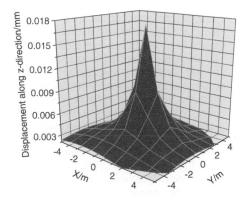

Figure 13.　Three-dimension graphs of displacements around the anchorage segment (z = 1.0 m).

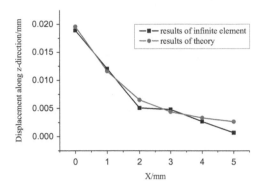

Figure 14. Contrast between the results of the infinite element and the theory (z = 1.0 m).

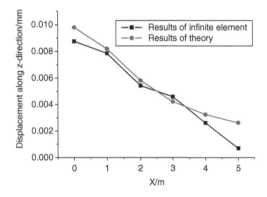

Figure 15. Contrast between the results of the infinite element and the theory (z = 2.0 m).

the results respectively computed by the infinite element method and the theoretic method derived by the author. The following characters of displacements of the rock around the anchor head can be obtained.

(i) Under the pre-stress action spot, the displacements of the rock around the anchor segment mostly focus on a zone center with the origin. The zoneís radius and depth are respectively 2 m and 4 m. In fact, the displacements above the action spot are similarity to the displacements under the action spot. Only the displacements under the action spot are press, and the displacements above the action spot are tension.

(ii) The displacements decrease with exponential forms. Thus, they form the displacement taper around the anchor head. Therefore, the stresses of the rock around the anchor head form a taper.

(iii) The results computed by the infinite element method are approximate to the results computed by the theoretic method derived by the author.

Only the curves of the infinite element method rebound. The reason is these elements are that the finite elements which will be likely to appear the phenomena.

3 CONCLUSIONS

From the analysis of the rock's displacements around the anchor head and segment, the following conclusions can be obtained:

(i) A compression stresses center region is generated in the rock around the anchor head by the pre-stress. Its radius is 2D (D is the side length of the anchorage pad pier) along the radial. In the region, the biggest values of the compression stress and the compression displacement are all at the center of the anchor. They decrease along the radial, and reduce along the axial when the depth increases.

(ii) The compression displacement and the compression stresses all decrease with exponential forms in the rock around the anchor head. Thus, their space shape are tapers taking the anchor as the axes.

(iii) The pre-stress generate stresses center zone in the rock around the anchor segment. Those stresses below the pre-stress action spot are compression, and those above the action of spot are tension. In the zone, the biggest values of the stress and the displacement are near the pre-stress action spot. These values decrease along the radial. At the same time, they decrease along the axial when the distance apart from the action spot increase.

(iv) The attenuation shape of the displacement and the stresses are all index near the anchor segment. Thus, their space shape are tapers taking the anchor as the axes.

(v) This paper introduces a new method coupling the infinite element method with the finite element method to study the anchor mechanism of stresses anchors. The obtained results have important role in the research of the anchor mechanism and engineering application.

REFERENCES

G. Beer and J.L. Meek, 1981. Infinite domain element. *Int. J. Number, Meth. Eng.* 17: 43–52.

Ge Xiu-run, Gu Xian-rong, Feng Bao-xiang, 1986. 3D infinite element and joint element. *Chinese Journal of Geotechnical Engineering*, 13(5): 51–56.

O.C. Zienkiewicz et al., 1983. A novel boundary infinite element. *Int. J. Number, Meth. Eng.* 19: 393–404.

P. Bettess, 1980. More on infinite element. *Int. J. Number, Meth. Eng.* 17: 1613–1626.

R.F. Ungless, 1973. *An infinite finite element.* M.A.Sc. Thesis, University of British Columbia.

Xiong Hui, He Yi-bin, Zhou Yin-sheng, 2005. A mapping infinite element with spatial whole matched type. *Journal of Hunan University (Natural Sciences)*, 12: 78–93.

Zhang Ai-jun, Xie Ding-yi, 2004. *3D numerical analysis of composite subgrade.* Beijing: Press of Science.

Zhang Jin-hua, Fu Li-xin, 2005. 3D infinite element computing modal for highway tunnels. *Central South Highway Engineering*, 3: 35–41.

Zhao Chong-bin, Zhang Chu-han, Zhang Guang-dou, 1986. Simulating the semi-infinite plane elastic ground by infinite element model. *Journal of Tsinghua University*, 5: 65–69.

Zhao Chong-bin, Zhang Chu-han, Zhang Guang-dou,1988. 3D fluctuate issue solving by the infinite element of the dynamical mapping. *Chinese Science*, 5: 59–63.

Underground Works under Special Conditions – Romana, Perucho, Olalla (eds)
© 2007 Taylor & Francis Group, London, ISBN 978-0-415-45028-7

Author index

Author index